Gourds in Your Garden

A Guidebook for the Home Gardener

Ginger Summit

Hillway Press
Los Altos, California

Author: Ginger Summit.
Illustrations by Ginger Summit.
Photographs by Ginger Summit unless otherwise noted.
Cover design: Kajun Design.
Book format: Graphic Express & Woodland Graphics.
Printed in Hong Kong.

Publisher's Cataloging-in-Publication Data

Summit, Ginger.
    Gourds in your garden: a guidebook for the home
gardener / by Ginger Summit
        p. cm
        Includes biographical references and index.
        ISBN: 0-9658691-4-8

1. Gourds 2. Gourd Craft. 3. Cookery (Vegetables)
I. Title
SB413.G6S86 1998   97-074384
635.9'3363   CIP
QB197-40939

Published by Hillway Press
PO Box 592
Los Altos, CA 94022

10 9 8 7 6 5 4 3 2 1

19-95

Acknowledgements

This book is the direct result of time and information that has been generously shared by so many members of the gourd community. Not only did they help me as I embarked on my own gourd garden adventures, but all provided a constant source of support and encouragement in the writing of this book. My special thanks go to *Jim Story*, for sharing his time and knowledge, as well as photographs and gourds from his garden. In addition, advice and suggestions were graciously offered by *Kern Ackerman, Suzanne Ashworth, Glenn Burkhalter, Bonnie Drake, Dr. Charles Heiser Jr.,* and *John and Jean McClintock.* Many thanks also to *Barbara and Hale Keller*, who read the many manuscripts of this work and helped clarify the myriad details leading to its publication.

Above all, I want to thank my husband *Roger* for his constant support in unraveling the mysteries and possibilities of computer technology. From the original research to the final printed page, he tirelessly and patiently guided me through the process of bringing a dream to reality.

CREDITS
Many thanks to the following individuals for sharing their photographs, their gourds and their gardens: *American Phytopathological Association, Helen Bos, Glenn Burkhalter, Barbara Keller, Jim Story, Jim Widess and Ardith Willner.*

Unless otherwise indicated, all photographs and illustrations were provided by the author.

Table of Contents

Introduction

When I first began growing gourds, I was a very enthusiastic gourd artist, surrounded by piles of seeds, armed with the little knowledge I was able to glean from general gardening books, naive and very optimistic. Literature assured me that even gourds that floated across the Atlantic still flourished on the shores of North America, unassisted by humans. I was further assured that gourds, after all, are cousins of zucchini, which I can grow with often too much success! I was also lulled into complacency by the fact that gardening books combine gourds in the same descriptive passages for growing all other melons and squash—if they are mentioned at all, that is! I learned a great deal that first year. My beginner's efforts were rewarded with some gourds, but not anywhere near the number and size I had dreamed about. I learned that these plants do require some special attention in order to grow healthy vines and produce satisfactory gourds.

Learning more about growing gourds has been a time-consuming task. They are largely ignored by most horticultural books and magazines, or are casually combined in sections that describe gardening techniques for the more popular cucurbit cousins, such as pumpkins and melons. Gardening techniques which focus on the qualities that make gourds unique in this huge vegetable family are largely ignored by the popular literature. The American Gourd Society does provide some pamphlets and brochures with valuable information, but this information is not readily available to the larger public.

I have written the book that I was searching for but could not find. It is intended for the home gardener who is enchanted with the quixotic shapes and sizes of this plant, including the crafter who is ultimately in search of the perfect gourd for a project, and the gardener who is fascinated with the beautiful lush vines and the wonderfully varied fruits.

Because the continental United States encompases a vast variety of growing conditions, the information in this book is necessarily general. However, it is intended to provide guidelines for the home gardener, suggestions for applying the information to individual growing conditions, and sources to look for specific help. Every locale will have specific gardening challenges, including such factors as soil, weather conditions, and space available for planting. In this book I have tried to present general horticultural guidelines for the three popular varieties of gourd being grown in the country today, and the special requirements that make these plants unique.

Details will vary throughout the country depending on a huge number of factors, and you the reader are encouraged to get more information from your local gardening center or county agricultural specialist. *Gourds in Your Garden* should arm you with enough information so that you at least know what questions to ask. The format was planned with wide margins specifically for you to note all of the supplemental details that are appropriate for your own unique growing conditions. This book is intended as a guideline to help you in the exciting adventures of growing gourds. If you find that details are not appropriate for your own conditions, I look forward to hearing from you. Even though gourds are one of the oldest cultivated plants in the world, they remain a mystery in the gardening literature. By sharing our experiences, we can begin to bridge the gap of the past thousands of years, as we once again learn to cultivate and enjoy one of Nature's greatest gifts to people.

Ginger Summit
Los Altos, CA
July 1997

This inviting arbor in the garden of Glen Burkhalter supports many Indonesian bottle gourds.

SECTION 1

What is a Gourd?
The Botanical Description

The simple question—What is a gourd?—unfortunately does not have a simple answer. When famed botanist L.H. Bailey wrote the book *Garden of Gourds* in 1956, he penned the following definition for gourds, which is still very much accepted in the general literature today: *"hard-shelled durable fruit grown for ornament, utensils and general interest."* Although gourds have been grown and used for many purposes other than those he mentions, the most important quality that we are interested in is the durable hard shell, which has been known to last for thousands of years. Undoubtedly it was the durable hard shell that first made the gourd significant to early humans as well, and encouraged its cultivation in locations literally covering the globe in all temperate and tropical climates where they could be grown. Unfortunately, the term 'gourd' has been freely applied not only to many other members of a very large plant family (Cucurbitaceae) which do not dry to a hard shell, and to the fruit of trees as well, which are totally unrelated to this botanical group.

The family *Cucurbitaceae* is one of the largest plant families, which has the following general characteristics:

- They grow on a vine.

- The vine is tendril bearing—that is, tendrils grow on the vine near the fruit.

- The leaves are usually five lobed, and grow alternately on the vine.

- The vines are monoecious—that is, the male (staminate) and female (pistillate) blossoms usually grow on the same vine.

- All cucurbits are plants which produce *fruits*—that is, they contain seeds and develop from an ovary inside the flower. The ovary of the female blossom is below the sepals and petals, usually a noticeable swelling that becomes the fruit if fertilization occurs. The fruit is identified botanically as a "pepo."

Within this large family are over 90 *genera* which share more specific characteristics. In the botanical lexicon genera are further sub-divided into *species*. A *species* is defined as a group of organisms able to exchange genes freely among themselves but not with other such groups. Each species has specific identifying characteristics which make it distinct from all other species within the family and genera. These characteristics are preserved because the many different varieties within the species may interbreed and create new hybrids or variations, but they will not mix or cross-fertilize with members of other species.

This book will focus on members of three genera: the lagenaria, the cucurbita pepo, and the luffa. The first two groups of fruits are most frequently referred to as gourds in North America, although most people don't realize they are related. The lagenarias are large fruits that dry with a hard woody shell, in a wide variety of shapes and sizes. The Cucurbita pepo var. ovifera, more commonly recognized as ornamental gourds, are familiar decorations in markets and homes during the fall, although most people don't realize that they too will dry with a hard shell, given proper care. The luffa also will dry with a hard shell, although most of the time that paper-thin shell is peeled away to reveal the more valued fibrous interior.

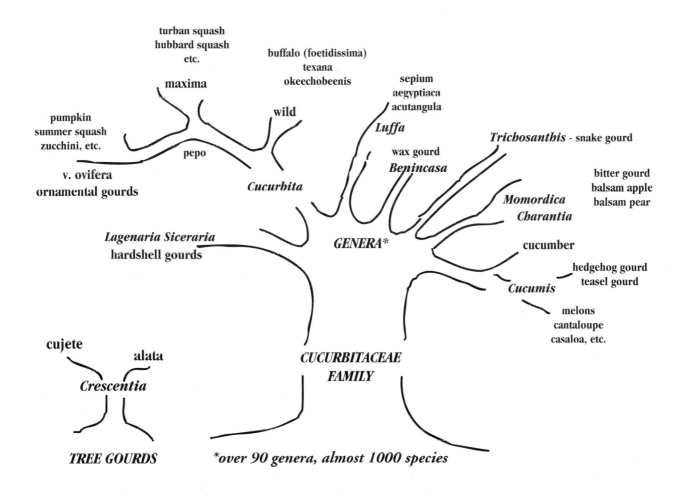

The diagram above attempts to relate some of the more popular members of the Cucurbit family, particularly the ones that are frequently called "gourds" in the popular literature. The family is actually extremely large, and only a tiny portion of the members are included here.

HARDSHELLED GOURDS

The large **hardshell gourd** is a member of the genus (singular form of genera) *Lagenaria.* (The name comes from the Latin *lagena*, meaning 'bottle'). It used to be referred to as *Lagenaria vulgaris*, but now is most often referred to in botanical literature as *Lagenaria siceraria* (from Latin *sicera*—drinking vessel)

This genus actually has six species, only one of which is commonly found outside of Africa. Within it are many varieties. Botanists and growers who are associated with groups trying to save historically significant seeds are attempting to cultivate these varieties in isolation in order to identify true 'types' which will reliably reproduce the same size and shape gourd in successive generations. Other growers continue to search for the true gourd varieties which maintain characteristics of the gourd as it was grown through the millinea in various parts of the world. However, because the gourd varieties interbreed so readily, it is very difficult to determine what is a true 'type' and what is a familiar hybrid.

Several attempts have been made to standardize the nomenclature for describing the various shapes and sizes of gourds. Names vary in different parts of the country and the world, which creates much confusion as crafters and growers describe their products. At this time no such standardization has been formally adopted by the American Gourd Society.

This list of hardshell gourds is by no means complete, but highlights the more familiar shapes the craftsperson may encounter. Keep in mind that most gourds are mixtures of these shape characteristics, thus creating individuals as distinctive as people. Also, size and shape of the gourd can be greatly affected by soil and weather conditions, as well as by the genetic inheritance contained within the seeds.

Cannon Ball Basket Ball

The following descriptions of some of the more popular shapes used by craftspeople include the terms recommended by the AGS in bold print, followed by some of the other locally familiar names:

Basket Type

These gourds are generally round, and either have no neck or are somewhat tapered toward the stem end of the body.

- **Cannon Ball** – perfectly round, ranging in size from tennis ball to slightly larger than a softball.

- **Basket Ball** – perfectly round, ranging in size from a small melon to a large basketball.

- **Canteen** – flattened round gourd, with the stem side flat or even slightly indented. Approximately 6-8" diameter.

- **Corsican Flat** – similar in shape to the Canteen gourd, but larger. This gourd can be up to 12-15" diameter.

- **Tobacco Box**, sugar bowl – flattened round gourd, with a slight taper at the stem end.

- **Bushel Basket** – round shaped body, although not as distinctively round as the basketball. May be slightly irregular in shape, but has no taper at the stem end. There is considerable range in size, but may exceed 20" diameter.

- **Kettle** – round gourd, larger than the bushel basket usually with a very thick shell. May be slightly tapered toward the stem end. These gourds, usually very large, can grow to be 20" diameter.

Canteen

Bushel Basket

Corsican Flat

Tobacco Box

photo: Jim Widess

Kettle

Sennari

Mexican Bottle

Miniture Bottle

Chinese Bottle

Birdhouse

Penquin

Indonesian Bottle

photo: Jim Widess

Wartie Hardshell

photo: Jim Widess

Lump-in-neck

Bottle Type

These gourds generally have a base bulb and a neck at the stem end. There may be a bulb in the neck, with a constriction or waist, separating the two bulbs.

- **Sennari** – these gourds are favored in Japan, where they have been extensively cultivated for sake containers. They have very distinct proportions, with a larger bulb at the bottom, a smaller upper bulb and a nipple at the top. The sennari gourd can range in size from miniature (2") up to 6" in height.

- **Mexican bottle** – similar in shape to the sennari gourd, but larger, 12" or more in height

- **Miniature bottle** – small dumb-bell shaped gourd with a larger bulb on the bottom, a smaller one on top. May range in size from 1" to 4" in height

- **Bottle**, **Chinese Bottle, Dumb-bell** - similar in shape to the miniature bottle, but may be up to 18" in height

- **Birdhouse** – similar to a small kettle, large bulb on the bottom, but a tapered stem end. May be up to 12" in height

- **Penguin**, **Powderhorn** – elongated bulb on the bottom, tapering up to a slightly narrowed neck. Usually up to 12" in height

- **Indonesian bottle** – large bulb on the bottom, with a smaller bulb on the top, separated by a relatively long slender waist

- **Lump-in-neck** – similar to an Indonesian bottle gourd and a dipper gourd, it has a lump in the neck with a narrow nipple at the stem end

- **Hardshell wartie** – usually similar in shape to a small birdhouse, but covered with warts, or bumps

Dipper Type

These gourds have a bulb at the blossom end of the gourd with a long thin neck extending to the stem end:

- **Short-handled dipper** – handle tends to be slightly thicker than the longer varieties—may be up to 8" in length

- **Long-handled dipper, extra long handled dipper** – these dippers can be grown to over 6 feet in overall length

- **Club, Cave Man's Club** – the bulb tends to be elongated with a gradual narrowing into the handle. Often the bulb merges into the handle gradually so that the entire gourd assumes the shape of a fat baseball bat.

- **Maranka** (Dinosaur, French Dolphin, Hercules Club, Alley Oop, Caveman's Club, Swan, Alligator) – this gourd differs from other dipper gourds in that the surface is marked by irregular ridges and grooves which cover the entire bulb area. Distinctive dark green when mature.

Short-handed Dipper

Long-handed Dipper

Club

Maranka

Grower Norm Brickner holding a prize-winning extra-long-handled dipper

Banana

Snake

Japanese long gourd

Trough/Siphon/Snake

These gourds are generally long with no distinct bulb.

- **Banana** – narrow banana-shaped gourd, generally narrower at stem end, up to 8" in length.

- **Longissima**, Baton, Snake – these long gourds can grow up to 4 feet in overall length. Note: There is a much narrower long gourd, also called the snake gourd, which is cultivated in Asia. It is of a different genus, *trichosanthes*, and does not dry with a hard shell.

- **Cucuzzi** – This snake-shaped gourd, also called the Italian Edible Gourd, is often raised as a vegetable. It will dry with a thin hard shell.

- **Zucca** – this trough shaped gourd can grow to huge proportions, up to 4 feet or more in length.

- **Japanese long gourd** – This is a long gourd, up to 6 feet, with a slight bulb at each end. Grown on a trellis, it will be straight but left on the ground it may assume interesting shapes.

Zucca

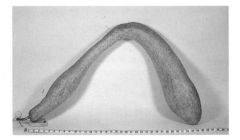

Japanese long gourd

16

ORNAMENTAL GOURDS: CUCURBITA PEPO

The *Cucurbita pepo* genus is one of the largest in the *Cucurbitaceae* family. It contains many of the edible relatives we are familiar with in food markets—some of the common varieties of pumpkins, many varieties of squash, as well as many other non-palatable fruits. While popular literature and local lore refer to many of these varieties as gourds,

especially in the autumn, the *cucurbita pepo-var. oviferis* is the only variety of this genus that dries with a woody hard shell, although that shell is often thin and somewhat fragile.

The Cucurbita pepo-v. oviferis, most commonly known as the **ornamental gourd**, contains many varieties, notable for their dramatic colors and shapes. The fruits are small, ranging from 1-6" in diameter. They are often called Mother Nature's showoffs, and they are readily found in markets and homes in fall decorations.

Ornamental Gourds come in a wide variety of shapes

Within the Ornamental variety, several sub-categories are used to identify the gourds further. These categories tend to be largely descriptive based on shape, size and texture: nest egg, orange, pear, spoon, Holy Crown (also called Crown of Thorns, Ten Commandments, or Finger), apple, bell, big bell, and depressa striata. Colors can vary within the categories from creamy white to yellow, orange, reddish orange, light and dark greens. These colors can be plain or striped, mottled or even bisect a gourd in half. The surface texture is also variable: they may be warty, grooved or smooth. The size of the gourd is often influenced by growing conditions of the plant as much as by variety, although generally ornamental gourds are under 6" in length or diameter, and are frequently much smaller than this.

Crown of Thorns always has ten fingers, which may be more or less pronounced

They also may be smooth or warty texture

The long Luffa cylindrica and ball-shaped Luffa sepium are more common in the United States

Luffa acutangula, with pronounced ridges, is a common vegetable in the Far East

LUFFA GOURD

The **luffa gourd** belongs to another genus, in which there are at least four species. The Luffa genus probably originated in the Orient where it was and still is grown largely as a food crop. The ridged species, the *Luffa acutangula*, is very common in Asian and Indian produce stores and farmers markets on the East and West coasts and throughout the South. If picked while young, it can be prepared much as other summer squash. The more mature fruit is quite bitter and has been used in folk medicine as a purgative. The *Luffa aegyptiaca* (sometimes referred to as *Luffa cylindrica*) is the smooth variety that is more often grown in the United States. When either variety of luffa is allowed to grow to maturity, the interior becomes a stringy, spongy mass. After it has been dried and cleaned, it is highly valued for a wide variety of uses. While you are probably familiar with the use of luffa as a bath or kitchen scrubber, it has also been used for mattress stuffing, hat lining (inside pith helmets or coolie hats in the Far East to provide ventilation), shoe lining, and during WWII as a filter for diesel oil in ships!

Luffas are frequently used by craftspeople and flower arrangers in many imaginative ways. Another species of luffa that is becoming popular with crafters is the *Luffa sepium* also known as *Luffa operculata* which produces a small round fruit approximately the size of a golf ball. While green the fruit is covered with spikes, but when it is dried and the shell is peeled off, the sponge is a small round ball.

Luffas require a slightly longer growing season to reach maturity, but they are not difficult to grow and are particularly well suited for trellises and arbors. Many attempts are being made to promote luffa gourds as a viable commercial crop in this country, since at this time most of the luffas used commercially in the United States are imported from Asia.

OTHER MEMBERS OF
THE CUCURBITA FAMILY
ALSO KNOWN AS GOURDS

To add to the confusion: In addition to the hardshell gourd (Lagenaria siceraria) and ornamental gourd (Cucurbita pepo var. ovifera), there are several other members of the Cucurbitaceae family which are referred to as gourds which do not necessarily dry with a hard woody shell. Many of these different types of gourds are cultivated and popular in specific geographical locations around the world and for very specific purposes. Because these plants are identified as 'gourds' in literature and seed catalogs, it is important to be aware of their existence. Some examples are:

Buffalo/Coyote gourds

- Buffalo gourd/coyote gourd

- Bitter melon/Balsam apple/Balsam pear

- Wax gourd

- Snake gourd

- Teasel gourd

- Hedgehog gourd

- Malabar melon

- Turk's Turban

   See Appendix III for a more detailed description of these other gourds.

Turk's Turban

Teasel gourd

Balsam apple

Malabar melon

Hedgehog gourd

Balsam pear

TREE GOURDS

Another fruit that is often called a gourd comes from a tree, not a vine. Two varieties of trees produce this fruit, the *Crescentia cujete*, and the *Crescentia alata*. Both grow in tropical areas in the Western Hemisphere, including Mexico, Central America, northern South America and the Caribbean Islands. The fruits suspend from the horizontal branches, woody globules sized from baseball to basketball, usually round or oval. When they are dried, opened and cleaned, their woody shell is used for many of the same functions as the hardshell gourds. The shell of the tree gourd is quite thin and very strong. These fruits are commonly used for bowls, utensils, floats, instruments, and frequently can be seen carved, painted or decorated much as other gourds, primarily for sale to tourists.

When the Europeans first began exploring and describing other lands and cultures in the 16th Century, they filled their notebooks with descriptions of the *calabaza* (Spanish) or *calebasse* (French). Both of these terms are derived from the Persian word meaning "melon." While in the Americas the term "calabash" is frequently used to describe the fruit of the tree, in Africa and South America "calabash" refers to all varieties of the hardshell gourds.

Cultural History

Evidence of gourds from ancient times has been found on nearly every continent, and dating from over 10,000 years ago. The story these remnants reveal, however, is still very incomplete.

While most of the earliest *archaeologic* evidence of gourds has been found in the Western Hemisphere, the *botanical* evidence suggests that the plant originated in Africa. Numerous wild relatives have been found on that continent which are possible progenitors of the modern hardshell gourd. (For a complete discussion of the possible origins of the gourd, please refer to the book by ethnobotanist Charles B. Heiser, Jr., *The Gourd Book*.) Botanist Thomas Whitacre conducted an experiment in 1954 in which he floated a bottle gourd in salt water for 347 days, and then stored the seeds for 6 years. After that time, many of the seeds were still able to germinate. This experiment suggests that it would indeed have been possible for a bottle gourd to have floated down African rivers and across the Atlantic ocean to land on various locations in the Western Hemisphere and throughout the world to other land masses. The many different locations where ancient gourd fragments have been found suggest that this may indeed have happened many times over the millennia, with gourds washing on shores from Brazil to Florida, India to China and beyond.

All of this speculation is focused on the hardshell gourd. However, many wild species of other members of the Cucurbitaceae family have been found in locations around the world, suggesting many different sites of origin for these diverse plants. The Luffa is probably originally from Asia. Many wild species of Cucurbita pepo are native to North America, hinting that this genus, or at least many of the species, originated in North America.

*Many instruments are made
of gourds in Africa today*

*Gourd containers are still frequently
used in villages in the Amazon basin*

Whatever the place or places of origin, it is abundantly clear that the gourd has been used by humans throughout the world for a very long time and for a great variety of purposes.

The oldest physical evidence of gourds come from the highlands of Peru, dating from 23,000-11,000 B.C. Along the coast of Peru, in the Huaca Prieta excavations, numerous remnants and even complete gourds have been found dating from 6000-4000 B.C.

In the Ocampo Caves in Mexico, gourd fragments were found dating from 7000-5000 B.C. In Florida, gourd seeds have been found in mastodon dung dated at 11,000 B.C. The oldest gourd fragments associated with human settlements in North America are also from Florida at the Windover site which was occupied around 7000-8000 years ago.

The Cucurbita pepo is one of the oldest domesticated plants, going back to at least 5000 B.C. in Mexico. Recent archaeological research suggests that the squash and hardshell gourd were being cultivated in the area of Missouri and Illinois by 5000 B.C. also, which would make it the oldest domesticated plant in North America. It is very possible that Cucurbita pepo was domesticated independently in both Mexico and North America.

Gourd fragments and seeds also appear along with evidence of earliest agriculture in many other parts of the world as well. In Africa, much newer, although still ancient, images, seeds and shards of gourds have been found in Egypt, Zambia and South Africa, dating from around 2000 B.C. Newer samples in other sections of the continent attest to the plant's widespread distribution and use. Prehistoric evidence of gourd usage has also been found in such diverse locations as Thailand, New Guinea, Hawaii and other Pacific islands. In Egyptian tombs, frescos show gourds being used as containers and utensils.

Gourds have been, and still are being used by cultures throughout the world for purposes as diverse as utensils, storage containers, musical instruments, objects of clothing and ceremonial and religious objects. Frequently medicine men or shamans have found special and sacred uses for the gourd, ranging from reliquaries to parts of a costume. In Hawaii gourds were used in every aspect of daily life by everyone from commoners to royalty, although only the gourds in the royal households were embellished with special decorations. (After European contact, gourds were gradually eliminated from those islands. Disease eradicated the plants, and the gourd utensils were replaced by more 'modern' containers of metal and pottery.)

When Europeans first arrived in North America, they found gourds being used for many purposes, including birdhouses. The Choctow and Chickasaw Indians found great advantage to attracting large colonies of purple martins to their villages. By the time the Europeans arrived, the martins were permanently conditioned to nesting in gourds erected close to the tribal living sites, and that dependence on human-provided homes continues to this day. Residents east of the Rocky Mountains are encouraged to erect purple martin homes, preferably made of gourds. Thus they are rewarded by a very effective bug control, and satisfaction that they are helping to maintain a healthy martin population. (See section on birdhouses page 100.)

Musical instruments made of gourds have been found in virtually every culture around the world in the temperate and tropical areas. It has been suggested that a gourd was probably one of the very first musical instruments, as it was picked up and shaken, and soon accompanied songs and ceremonies. The gourd shape is a natural resonator, and was soon incorporated into drums and many other forms of instruments in which the sound needed additional amplification. Stringed instruments probably originated in Africa, where they still assume a wide range of styles, including harps, lyres, and progenitors of the guitar and banjo. The stringed instrument was elaborated in India, where two gourds act as resonators on the beautiful sitars. Gourds have also been used for whistles, horns and flutes, including the unusual nose flutes still in use in many island cultures in the Pacific. Most of these instruments, while modified, are still in use in many forms today throughout the world.

These popular instruments are frequently heard in the Caribbean and South America

This gourd from Colombia was pierced with holes to function as a sieve

*Sherlock Holmes smoked
a calabash pipe*

*These containers from Japan
and Europe were both designed
to hold wines*

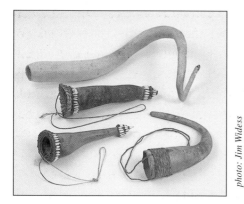

*Penis sheaths are still worn in the
highlands of New Guinea*

Further evidence of the importance of gourds in early cultures is the prominence of that plant in mythology and folk tales around the world. Gourds were often thought to be an intermediary between the visible world of humans and the invisible spirit world, both in ceremony and legend.

While gourds were gradually replaced in most modern cultures by other materials, they continue to be an indispensable household tool in many parts of the world. As people are becoming more interested in exploring their social and cultural origins, gourds are once again emerging as a fascinating and useful plant that fits many diverse needs in people's lives. Artists, craftspeople, musicians and historians continue to explore old and new ways to utilize this most flexible and adaptable of plants. Once again gourds are assuming an important role in our daily lives.

photo: Jim Widess

*Lime containers
made of gourds are
frequently used in
South America and
New Guinea*

Fertility Gourd from Zaire

SECTION II

The Gourd Garden

Because gourds were so fundamentally important in the lives of most prehistoric people, it is little wonder that many cultures established certain rituals or ceremonies around something as important as their planting in the spring. These rituals reflect the cultures they were part of— usually the individuals who wielded the greatest power were the ones selected to plant the sacred seeds:

In New Zealand, the man who was selected for this important task held a seed between thumb and finger, faced the east and raised his outstretched hands above his head to make a large circle as he sang the ceremonial songs. This supposedly instructed the seeds to assume certain shapes.

In Hawaii, the gourds were planted and tended only by pot-bellied men. Before actually planting, he ate a full meal to expand his body even further into the desired shape of the gourd. Then, huffing and puffing, he struggled out to the field, dug a hole, and dropped in a seed, singing a song petitioning the gods to grant a bountiful harvest. Only men were allowed to cultivate and eventually harvest the crops in this special garden.

Women were specifically prohibited from gourd cultivation in Europe as well. In a gardening handbook published in the sixteenth century in England, *The Gardener's Labyrinth*, it was cautioned: "This one matter ought especially to be cared for, as Columella after the Greek Florentinus warneth, that no woman come or very seldome approach nigh to the fruits of the Gourdes or Cucumbers, for by her onely handling of them, they feeble and wither, which matter if it shall happen in the time of the Termes, doth either then slay the young tender fruits with her looke, or causeth them to be unsaverie, and spotted or corrupted within."

In West Africa some tribes specifically designated gourd cultivation as a woman's task. (Many of these tribes were matrilineal, so women held positions of great power in the villages.) Because gourds were so important in all household activities, gourd containers were not only prized possessions to be passed from generation to generation, but sometimes actually comprised a woman's display of wealth.

In remarkable contrast to these elaborate ceremonies, Native American tribes in the vicinity of what is now Virginia felt it was unlucky to intentionally plant a gourd seed. In these tribes the individual, usually a man, walked out to the field feigning indifference, and casually cast the seeds over his shoulder. By good fortune, enough of the seeds broadcast in this manner flourished into plants bearing fruit that gourds were able to continue as an important part of the village life.

Today we use many other techniques in the spring as we plant our gourd gardens. But our hopes are the same as the ancestors hundreds of years ago a successful crop of these wondrous fruits.

Preparation

One of the most fascinating aspects of gourd cultivation is that it will provide satisfaction for gardeners with any level of experience. The beginning gardener, armed with more dreams than experience, will enjoy a fascinating crop. Additional knowledge and effort reward the gardener with not only more abundant crop but lots of opportunities to experiment with crop manipulation at every level—cross pollination, training, and raising exotic fruits.

Gourds can adapt to a wide variety of growing conditions. Small varieties, especially the ornamental gourds, are highly suitable to container gardening. Because all gourd vines like to climb, problem areas in the garden or yard can be beautifully disguised under a blanket of thick foliage and abundant flowers. Although gourd vines are notorious space grabbers, they can be trained and confined to a fence, trellis, or other frame.

Three principal factors influence the growth of healthy gourd plants: 1) plenty of sun, 2) well fertilized soil, and 3) lots of water.

SELECTING THE GARDEN SITE

In selecting an appropriate site for the gourd garden, look for an area that gets full sun. Gourds require 120–140 frost-free days to mature. This translates to at least 4 months between the last spring frost and the first frost of the fall. (See chart on pgs. 28 – 29.)

In appendix I is a map of the United States with growing zones identified, developed by the U.S. Department of Agriculture. It is based on winter minimum temperatures and is useful if taken as a very general guideline. However, it does not take into account summer high temperatures, rainfall patterns, and other factors that have important influence on growing season. Each locale has microclimates that will greatly affect when plants should be started and where they should be placed to take maximum advantage of the environmental features. Check with your local nursery for specific information about your particular setting.

If you live in an area that has a short growing season or fewer than 120 days between frosts, it is possible to start the plants indoors early, and then transplant them when the ground is warm. Keep in mind that plants may suffer a transplant shock when they are put in the ground, sometimes as much as two weeks. Many growers like to start their plants indoors even when they have an adequate growing season, as a way to prevent some damage to the baby plants by garden pests (see pp. 48).

The soil can also be warmed by covering it with black plastic prior to planting. Several types of garden plastic or horticultural mesh are available in garden supply stores and may be a good solution to extending a short growing season. Row covers over baby plants further extend the season.

It is important, once the plants are in the ground, that they get full sun, or at least sun for six hours a day. (However, if you live in an area that gets intense sun in the summer months, the plants may appreciate a filtered sun for at least part of the day. Some of my plant vines extend to the protection of partial shade of trees. They do not wilt in the heat of the day, and seem to grow just as vigorously as those in continuous sun.)

Next decide on the planting layout that is best for the garden space available and the gourd plant. Gourds can be planted in either rows or in hillocks: the most important consideration is to be sure to allow adequate room for both roots and vines to spread. Recommended spacing is 8 feet between hills or rows, and a minimum of 4 feet between plants grown in rows. (Many commercial growers recommend at least 8-10 feet between the plants in all directions. However, many home gardens do not enjoy this luxury of space. I have grown gourds successfully leaving approximately 4 feet between plants.) When other vegetables or plants are grown in the same area as gourds, be sure the gourds are on the periphery of the garden plot so they can spread without smothering the other plants.

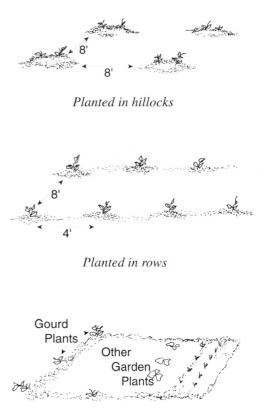

Planted in hillocks

Planted in rows

Planted in a garden

FROST DATA FOR SELECTED LOCATIONS

Location	LAST FROST		FIRST FROST		FROST-FREE DAYS	
	Average	90% Safe	10% Chance	Average	Average	90% Safe
Albany, NY	May 7	May 24	Sept. 19	Sept. 29	144	126
Albuquerque, NM	May 5	May 26	Sept. 25	Oct. 9	159	130
Anchorage, AK	May 4	May 15	Sept. 10	Sept. 25	143	125
Asheville, NC	Apr. 10	Apr. 24	Oct. 11	Oct. 24	196	179
Augusta, GA	Mar. 28	Apr. 15	Oct. 23	Nov. 6	222	199
Baltimore, MD	Apr. 11	May 4	Oct. 8	Oct. 21	186	161
Billings, MT	May 12	May 29	Sept. 6	Sept. 23	133	110
Birmingham, AL	Mar. 31	Apr. 16	Oct. 24	Nov. 6	221	202
Bismark, ND	May 14	May 26	Sept. 7	Sept. 20	129	111
Boston, MA	Apr. 26	May 12	Sept. 29	Oct. 11	168	149
Burlington, VT	May 11	May 25	Sept. 19	Oct. 1	142	123
Charleston, SC	Mar. 18	Apr. 6	Oct. 30	Nov. 12	239	217
Chicago, IL	Apr. 22	May 6	Oct. 10	Oct. 23	187	165
Cleveland, OH	May 4	May 18	Oct. 5	Oct. 18	167	147
Columbus, OH	Apr. 26	May 9	Oct. 3	Oct. 17	173	156
Dallas, TX	Mar. 25	Apr. 11	Oct. 26	Nov. 12	231	210
Denver, CO	May 3	May 20	Sept. 20	Oct. 8	157	131
Des Moines, IA	Apr. 28	May 11	Sept. 24	Oct. 7	162	144
Detroit, MI	Apr. 27	May 12	Oct. 5	Oct. 19	175	153
Fairbanks, AK	May 17	May 25	Aug. 24	Sept. 5	112	94
Gainesville, GA	Apr. 3	Apr. 18	Oct. 23	Nov. 6	216	198
Galveston, TX	Jan. 28	Feb. 25	Dec. 10	Jan. 9	345	308
Honolulu, HI	—	—	—	—	365	365
Jacksonville, FL	Feb. 14	Mar. 14	Nov. 16	Dec. 14	303	267
Kansas City, KS	Apr. 18	May 2	Oct. 3	Oct. 18	175	158
Las Vegas, NV	Mar. 7	Apr. 3	Nov. 7	Nov. 21	259	231
Little Rock, AR	Mar. 25	Apr. 9	Oct. 24	Nov. 5	234	212
Los Angeles, CA	—	Feb. 3	Jan. 1	—	365	350
Louisville, KY	Apr. 20	May 6	Oct. 6	Oct. 20	168	150

Location	LAST FROST		FIRST FROST		FROST-FREE DAYS	
	Average	90% Safe	10% Chance	Average	Average	90% Safe
Memphis, TN	Mar. 23	Apr. 8	Oct. 27	Nov. 7	228	207
Miami, FL	—	—	—	—	365	365
Milwaukee, WI	May 4	May 18	Sept. 26	Oct. 11	156	132
Minneapolis, MN	May 9	May 25	Sept. 13	Sept. 25	138	117
Mobile, AL	Feb. 27	Mar. 19	Nov. 5	Nov. 26	272	241
Nashville, TN	Apr. 5	Apr. 16	Oct. 14	Oct. 29	207	189
New York, NY	Apr. 1	Apr. 13	Oct. 27	Nov. 11	223	203
New Orleans, LA	Feb. 20	Mar. 21	Nov. 15	Dec. 5	288	249
Norfolk, VA	Mar. 23	Apr. 6	Oct. 31	Nov. 17	239	216
Oklahoma City, OK	Apr. 4	Apr. 17	Oct. 18	Oct. 31	208	189
Omaha, NE	Apr. 27	May 11	Sept. 25	Oct. 9	163	143
Orlando, FL	Jan. 24	Mar. 4	Dec. 4	Jan. 3	343	249
Philadelphia, PA	Apr. 3	Apr. 14	Oct. 28	Nov. 13	224	201
Phoenix, AZ	Feb. 5	Mar. 16	Nov. 18	Dec. 15	308	265
Pittsburgh, PA	May 1	May 18	Sept. 28	Oct. 13	162	139
Portland, ME	May 10	May 25	Sept. 18	Sept. 30	143	121
Portland, OR	Apr. 3	Apr. 26	Oct. 18	Nov. 7	217	187
Raleigh, NC	Apr. 11	Apr. 29	Oct. 16	Oct. 27	193	176
Richmond, VA	Apr. 10	Apr. 27	Oct. 13	Oct. 26	198	178
Rochester, NY	May 3	May 18	Sept. 29	Oct. 15	164	140
Salt Lake City, UT	May 10	May 27	Sept. 23	Oct. 7	150	127
San Antonio, TX	Mar. 3	Mar. 23	Nov. 6	Nov. 24	265	235
San Francisco, CA	Jan. 8	Feb. 9	Dec. 8	Jan. 5	365	340
Seattle, WA	Mar. 24	Apr. 20	Oct. 27	Nov. 11	232	198
Spokane, WA	May 4	May 20	Sept. 19	Oct. 5	153	131
St. Louis, MO	Apr. 16	May 1	Oct. 6	Oct. 21	188	163
Topeka, KS	Apr. 21	May 4	Oct. 1	Oct. 14	175	158

NOTE: These figures are appropriate for areas immediately surrounding reporting stations. Variations within small geographic areas can be significant, particularly near mountains, oceans, and large bodies of water. Consult the National Climatic Data Center, Asheville, NC, for data on other specific locations. Keep in mind that this information reports data pertaining to frost temperatures; it does not indicate high temperatures that are also important for gourd growth.

Many types of gourds develop best if grown on a trellis or above ground. There may be a fence or wall that can serve as an appropriate support (see pp. 39) In this case, cultivate the soil at the base of the support (wall or fence) to a distance of at least 4 feet to insure the soil is loose, fertilized, and can support a vigorous vine.

It is also possible to grow gourds in containers. One important advantage is that containers can be moved to sunny or protected areas to insure plenty of sun and warmth. In this case be sure to select gourd varieties that are small, such as ornamentals, mini-bottles, sennari, mini-dippers, etc. Fill the container with well mixed soil that has good drainage. Several types of supports can be designed for a container that will allow room for the vine to climb and spread (see section on trellises and supports).

GOURD VARIETIES SUITABLE FOR CONTAINER/TRELLIS/GROUND

Containers	Trellis	Ground
all ornamentals	luffa	bushel types
sennari	bottle	African
mini-bottles	Indonesian	bushel basket
	lump-in-neck	kettle
	canteen	zucca
	all dipper types	all the types listed
	maranka	in the other categories
	snake types	
	birdhouse	
	penguin	

WELL FERTILIZED SOILS

Gourds require well fertilized, rich soil. However, before applying additives, it is a good idea to have the soil content analyzed for such factors as pH balance and mineral content. Test kits are available in most nurseries either to test the pH levels yourself, or send soil samples to a lab which will analyze them and then send you a report with recommendations. It is also possible to take a soil sample to the local agricultural extension office to have it tested. (The phone book should have this listed under State or County government listings.)

pH SOIL CONDITIONS

1 - - - - - - - - - - - - - - 6.0 - 6.5 - - 7 - - - - - - - - - - - - - - - - - - 14
totally acidic   ▲   neutral   totally alkaline

**ideal gourd
growing**

more common in areas   more common in areas
with lots of rain and   with light rainfall
sandy or loamy soils   chalky soils
peaty woodland soils

To raise pH levels, add: To reduce pH levels, add:
  lime   sulfur
  wood ashes   peat moss
  organic matter   evergreen needles
  seaweed, calcified   ground bark

PH balance refers to the acid/alkaline balance of the soil. The pH scale goes from 1 to 14, with 1 being totally acid and 14 being totally alkaline. Neutral soils measure at 7.0. Gourds like soil that is slightly acidic, that is between 6.0 and 6.5 Very acidic soil may require the addition of lime to raise the pH levels; if the soil is too alkaline, it may be necessary to add sulfate of ammonia to reduce the pH.

Some people add ashes or lime to deter certain garden pests such as snails or grubs and larva of other insects. You should be aware that these additions will affect pH, so be familiar with the basic composition of the soil before starting to add additional elements to it. Your nursery will be able to recommend other steps to take to insure an appropriate balance for a vegetable garden.

In addition to pH balance, the three important components of the soil which will be tested are nitrogen, phosphorus, and potassium. Nitrogen is the basic building block of protein and is important to healthy stem and leaf production. Nitrogen is not a mineral, but comes from organic matter, fertilizers or some groundwater. Plenty of nitrogen is important initially in the growing season to insure strong vine growth. Phosphorus stimulates root development in young plants; when applied later in the growing season, it encourages a greater proportion of fruits per plant. Potassium is critical in the formation of starches, sugars and cellulose. As the plants mature, it also may help build resistance to fungal diseases. Fertilizers generally indicate on their packaging three numbers, which refer to the balance among the nitrogen, phosphorus and potassium.

The information provided by the soil tests may indicate other important measurements of soil composition, such as magnesium and calcium. Minute quantities of these minerals are required in varying amounts depending on the plants. Trace amounts are usually included in most all-purpose fertilizers, so normally you do not have to add additional minerals to the soil.

Many experienced farmers begin to prepare their soil as soon as they harvest the last crop in the fall. This is a good idea for the home gardener as well, for you can eliminate or reduce serious pest or disease problems by maintaining a good garden routine. The first step is to remove completely the old vines and any old leaves or fruit that may harbor insects, their eggs or larvae. Planting a cover crop such as vetch, fava beans, clover or mustard will replace some of the nitrogen content of the soil, and provide carbon and other organic material for the following season. Plant these crops in October or early November, and then till them back into the soil approximately one month before planting in the spring. You can also top-dress the soil in the fall with cow manure and bone meal. These help to replace the phosphorus and potassium.

Weather permitting in the spring, the ground should be prepared well in advance of planting. If the soil is shallow or has poor drainage, at least make special preparation in the immediate areas where the seeds will be planted: dig a hole up to three feet deep and three feet wide and fill with a mixture of soil, compost and manure.

In preparing the soil, add manure and a good well balanced fertilizer (10-10-10: that is, equal balance of nitrogen, phosphorus and potassium) far enough in advance of planting so as to allow the manure to decompose and not burn the seedlings.

Additional Fertilizers

If you prepare the ground thoroughly before planting the gourds, you may not have to add much fertilizer during the growing season. Manure that is obtained directly from the source, that is, from a farm or other location that keeps animals, is very rich indeed, and will greatly enrich the soil. Some manures and fertilizers that are sold for the home gardener have been diluted with inert materials. While these are still good for loosening the soil, they do not add the quantity of active ingredients which you may have anticipated. Therefore, check the labels for the percentage of active ingredients before spreading them on the soil. You can add a fish emulsion or general purpose 10-10-10 (equal parts nitrogen, phosphorus and potassium) as the plants begin to vine. Once the blossoms appear and the baby gourds begin to set, however, do not add any more nitrogen to the soil. Promoting vigorous vine and leaf growth will take energy and nutrients away from the fruits. Most nurseries and garden centers carry fertilizers containing higher amounts of potassium and phosphorus which are designed specifically to promote blooms or fruiting.

Some gourd growers like to provide their special plants frequent doses of "manure tea." To make this concoction, put some manure in a five gallon container and fill with water. (Some recommend suspending the manure in a cloth bag, much like a giant tea bag. However, manure in the bottom of a container works just as well.) Stir the solution to make sure the nutrients are suspended in the water, and dilute to a 'weak tea' consistency. Pour a bucket of this "tea" on your plants frequently. Then stand back and watch the gourds swell with appreciation!

Usually gourd growers are interested in growing gourds that are especially long or large. However, some people prefer to have very small gourds. In this case, it is possible to stunt the size of the fruit by planting in poor soil and providing just enough water to maintain the vines. Gourds have thrived throughout the world in vastly differing climates, soils and growing conditions. If you do not have the perfect garden soil, don't despair. Improved soil will increase the size and number of the crop, but even if the soil is less than perfect, you will still be rewarded with many fascinating gourds.

PLENTY OF WATER

Gourds require regular watering throughout the growing season, especially right after the seeds or seedlings are planted. Many irrigation systems are available for the home gardener which can be laid out in advance to insure a regular and consistent watering schedule. Drip systems and soaker hoses are excellent choices, since they deliver water directly to the soil and don't wet the leaves. Wet leaves of gourd plants are susceptible to mildew and other fungal problems; therefore, overhead watering systems are not recommended.

One handy way to irrigate seedlings is to surround the hillocks in which they are planted with an old tire. Cut the tire in half around the circumference to create a circular u-shaped reservoir. Drill holes in the sidewalls to allow the water to drip evenly and soak the seedlings without having to irrigate an entire field.

How Much Water?

At maturity, gourds, like many of their squash and melon cousins, are 90 95% water. The plants, therefore, are extremely thirsty throughout the growing season, since much of the water is directly translated into fruit composition. Also, the gourd vines are prolific and the leaves are often very large, the size of dinner plates! Much moisture is lost through evaporation from all this leaf surface.

The watering schedule will vary in different parts of the country, depending on rainfall, humidity, soil drainage, temperature, etc. New seedlings should be kept moist until the roots are firmly established, or approximately four-six weeks. Then the water schedule can be reduced to the point that healthy vine and fruit growth can be maintained. Usually this will be about one or two times per week.

During the hot summer sun, the leaves of the hardshell gourd often wilt and droop. Watch to see if the leaves recover as the sun goes down. If not, you may need to increase the water schedule. Also it is important to keep in mind that the roots are shallow and can cover great distances, like the vines above the ground. Make sure the watering system takes this growth pattern into effect.

In the early fall the watering schedule can be decreased or stopped, to allow the gourds to begin "hardening off." As the vines wither away, your precious crop will remain as proud sentinels in the field, rewards of a season of labor.

*A thick protective cover
of mulch is spread under this arbor.
These prolific vines
belong to Jim Story.*

*This well mulched seedling
is in the garden of
Mr. Miyajima, Matto, Japan.*

Mulch, the Gardener's Best Friend

Mulch can be used to solve many of the problems faced by the gourd gardener. When spread over the ground in early spring, mulch will retard or eliminate pesky weeds. By spreading a generous layer of mulch between the rows or hillocks of plants, weeds will either be smothered under the layers or will have such shallow roots they will be easy to pull. If you put down your watering system before you add the layers of mulch, the evaporation will be minimal and the leaves of the plants will be protected from the frequent water. As the gourds grow, they will have a nice soft and dry bed to rest on as they mature. At the end of the growing season, the mulch can then be tilled into the ground in the fall after the crop has been harvested. This adds important organic material to the soil.

There are several different types of mulch you can consider. Some examples are: straw, grass clippings, shredded leaves, and wood chips. Some gardeners have even used shredded newspapers. Mulches are inexpensive and considering all their benefits, should be an essential ingredient in every garden plan.

*Work mulch
around young plants.*

PLANTING OPTIONS TO CONSIDER

Because gourd vines are aggressive and prolific, many gardeners find that it is appropriate for their own garden spaces to train gourd vines on vertical supports instead of allowing them to trail on the ground. There are many advantages to growing gourd vines upright:

- They are natural climbers, and frequently climb trees, posts and buildings even when they have not been trained to do so.

- The plant, with its large profuse leaves and lacy white or yellow blossoms, is extremely decorative. It makes a beautiful addition to a landscape, such as hiding an unsightly view, fence or shed.

- Gourd leaves should not be wetted unnecessarily. By keeping the vines off the ground, they are safely removed from irrigation, ground pests and diseases as well.

- Many gourds, such as the luffa and dipper gourds, grow much better suspended from a trellis. Other small and medium-sized gourds are equally suitable for trellising: their shapes tend to be more regular, they do not discolor on one surface where they are in contact with the ground, and they are safely removed from many pests such as snails and slugs that like to graze on the gourd epidermis.

For large gourds, a sling can be made from pantyhose.

CAUTION: Large heavy gourds should not be grown on most trellises. Tendrils and vines cannot support the cumulative weight of many growing gourds. Heavy unsupported gourds occasionally crash to the ground, breaking both vine and trellis. Large gourds can be grown on specially designed, sturdy trellises, as long as they are individually supported to take much of the weight off the vine itself. (These supports can be easily fashioned out of panty hose, or net bags such as those produce is packaged in. They should be soft and flexible enough to allow unrestricted growth of the gourd, but strong enough to support the weight of the mature fruit.)

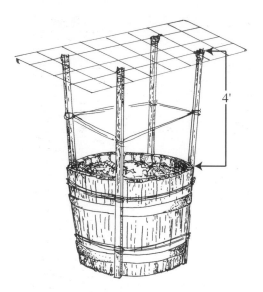

*Tie four supports
around an oak barrel*

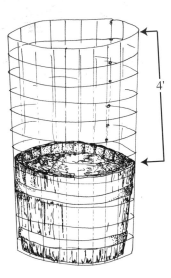

*To make an easy trellis for
a gourd container, encircle
it with a wire cylinder*

A second minor drawback to growing gourds on a trellis is that the vine is not able to re-root along the length of the vine. This is especially a problem if a squash vine borer attacks the plant. Vines growing on the ground can easily be re-rooted at a joint where the leaves and tendrils emerge, thus saving the vine. When grown on a trellis, the vine attacked by a squash vine borer will most likely die.

CONTAINER GARDENING

If gourds are grown in pots, a trellis can be constructed around or in the pot. This allows the vine to grow upright instead of hang over the side. Many design ideas for trellises can help to create an attractive centerpiece for your garden.

- The simplest structure to make is a cylinder of medium gauge wire mesh which can be wrapped around the pot and extend approximately 4 feet above the upper edge of the pot. Poles can be pushed into the container around the perimeter in 3-4 places, which then provide additional support for the wire mesh. As the gourd plants begin to vine, carefully direct them to the wire support. Once they begin to climb the wire mesh, they will continue climbing upward and can be tied occasionally.

- Another type of trellis for a pot can be constructed of four wood poles, stuck into the container around the perimeter of the pot and extending at least 4' above the top rim. Twine laced between the poles provides additional reinforcement and gives the vines more to grasp onto.

- PVC pipe can be used to create a very simple trellis in a pot.

- Another very simple support can be created by sticking a single rod or pole down the center of the pot. Heavy garden twine can be tied at the top of the pole and brought down to the rim of the pot. (I made a ring of several thicknesses of twine which I tied secure around the rim of the pot. The separate pieces of twine coming from the top of the pole were then anchored to this ring, creating a teepee effect with the twine.)

• Many commercial lattices are available for the container gardener. While they are attractive, many are constructed with staples or glue which simply are not strong enough to support a gourd vine. These trellises may be used if all joints are reinforced first with wire or through-hole bolts to prevent a sudden collapse in the middle of the gourd growing season.

TRELLISES AND ARBORS

When planning a trellis for the garden, there are many factors to keep in mind:

Mini-gourds are suitable plants for a container on the patio

• **Permanent or temporary:** Home gardeners may want to design a support system that they can remove after the growing season, or move from one area to another. Gourds do deplete the soil of many nutrients, and also are susceptible to garden pests that become exaggerated if they are grown in the same spot year after year. Trellises can be designed that are both strong and temporary, and therefore can be moved to different locations in the garden.

• **Strength:** Gourd vines are prolific growers. They have been known to reach up to 100 feet in length, although more frequently they stretch 30-40 feet. The main stem is strong and flexible, but slightly woody, and the vine is supported by tendrils that grasp tenaciously around any possible support. The cumulative weight of the gourd vine and the mature fruits can be very very heavy. Therefore, you must plan a system that can bear not only weight but the twisting and torque that can be exerted by the growing vines.

• **Dimensions:** The type of gourd you are growing will help determine the size requirements of the trellis. A long handled or extra long handled dipper requires a taller structure than ornamental gourds or mini-bottles, for example. Many growers like to create long low trellises, just high enough to keep smaller gourds off the ground.

Plant gourds along a cyclone fence to hide an unsightly view.

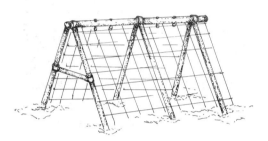

Cover an outgrown child's swing with wire mesh.

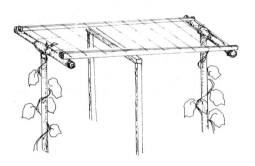

Use old clotheslines to make a trellis. Reinforce with wood supports, and cover top with wire mesh.

• **Climate:** Keep in mind the weather patterns in your area when planning a gourd arbor. Supports that are appropriate in a mild climate, such as California, may be completely inappropriate in an area that experiences severe summer storms or heavy winds. While the structure is perfectly capable of supporting the vines and gourds, it may collapse in certain weather conditions. Prepare for the extremes, so that you are not threatened with a total collapse in the middle of the growing season.

Using supports already in place:

• Most fences are readily adaptable to support a beautiful crop of gourds. The main requirement is that the plants receive full sun. The best kind of fence is a wire mesh, or cyclone type, so that the vine has unlimited places to grip and support itself. If your fence is solid, such as wood or stucco, you may want to face it with chicken wire for the growing season, or else attach wire supports for the vine to grab. Always keep in mind that the weight of the gourds will provide quite a stress on whatever support you attach.

• A child's climbing apparatus (that has been outgrown, hopefully!) is an ideal trellis for gourds. It is usually sturdy, has many legs and can support several vines. A child who has outgrown the swing and slide stage may enjoy his recycled play area as a secret hideaway with wonderful mysterious shapes hanging in all directions.

• Old clotheslines are usually placed in a sunny area, and often can be reinforced in such a way to support a nice crop of gourds. Stretch heavy gauge wire mesh such as concrete reinforcing wire over the top of the hanging lines to provide additional support for the gourds.

Temporary: (movable)

- Ladders: If you can find old wooden ladders, especially the tall ladders once used in orchards, they can be easily adapted to many types of trellis structures. Opened, they form a ready-made A frame. Attaching wooden 2" x 2" poles at the top on either side extending down to the ground will strengthen the balance of the ladder and create a tent-like structure. Use twine or fruit netting between the sides of the ladder and the poles to provide additional support for the structure.

- Put several ladders in a row and connect them with wire mesh or twine to create a support for gourds which have been planted in rows.

- A teepee of sturdy poles can be made to any dimension that will fit in the garden. Again, the most important consideration is the type of gourd you are growing. A small teepee will become a flowering highlight of the garden or patio in the summer and can bear a prolific crop of mini-gourds and ornamentals. A larger teepee with poles 8 - 12 feet, can be expanded to cover a larger diameter, support more plants, and allow dippers or other gourds to hang down in the center. Be sure to allow for the weight of the mature fruits by using sturdy poles. Join the poles laterally with wire mesh, both to give the vines support and also to help balance the teepee.

- PVC pipe comes in a variety of weights, and with a wonderful assortment of joiners. A structure can be created to adapt to almost any garden requirement. Be sure to use the heavy-duty pipe to support the weight of heavier gourds. For the ornamental and mini-bottles, the lighter weight pipe is sufficient.

You may want to use a special anchor for the bottom of the PVC pipes. I pounded 4' lengths of re-bar into the ground, and then slipped the PVC pipe over it to create a very solid support.

Make a teepee out of sturdy timbers to provide a flexible support in the garden

Heavy-duty PVC pipe can be fashioned into many different arbor designs. Cover with wire mesh to provide extra support for the pipe and for the vines.

This wooden trellis can be moved each season.

This old orchard ladder has been given new life as a decorative patio arbor. (Garden of Helen Bos)

- Movable Wooden Trellis: An exceptionally strong and flexible trellis can be created with 2"x 4"s and concrete reinforcing wires:

Make a square with the 2"x 4"s reinforcing the corners with metal strips. Most 2"x 4"s come in eight foot lengths, and will make an eight foot square, a nice flexible size for many gardens. However, the 2"x 4"s frames can be cut and made to any dimension to fit your garden space. After the squares have been created, secure the reinforcing wire in place with heavy duty staples. A single frame can be braced against a wall or solid fence to create a lean-to support for vines. It can be pitched at any angle that fits the garden space, or even can be mounted flush with the wall of a house or garage.

Two frames can be hinged together at the top to create an A-frame structure. This can be stood between two rows of gourds, or placed so that the legs or outside corners of the frame are positioned between four hillocks, to provide ample growing room for several vines.

Permanent Structures:

If you are a true gourd lover and know that you will be growing gourds for many years, consider building a permanent structure. Select materials that are very sturdy to withstand the weight of many seasons of large gourds. Most permanent trellises are constructed of 4"x 4"s or material of similar dimension.

Space the sides of the trellis at least 6 to 8 feet apart, or wide enough to accomodate any equipment that may be used during the year, such as rototiller or small tractor. The lateral supports of the sides can be placed anywhere from 4 to 8 feet apart, depending on the materials used and the size of the overall trellis. The trellis should be tall enough so that an adult can easily walk underneath, but low enough to reach through and handle the vines or gourds growing across the top. The entire structure should be reinforced with heavy gauge wire mesh. Concrete reinforcing wire mesh with 6 inch squares is ideal for this purpose, since it is very strong, and yet has large enough openings for the small gourds to hang through.

You may want to plan an irrigation or watering system as part of the structure. A drip system that waters only the soil and roots is much better than an overhead sprinkler system. Gourd leaves are particularly susceptible to mildew, and they should not be wet any more than necessary.

Wooden frames covered with concrete reinforcing wire are strong supports for the gourd vines

By mid-summer, the trellis is completely hidden by profuse gourd vines. (Garden Photos: Glen Burkhalter)

Planting The Garden

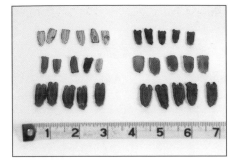

Hardshell gourd seeds come in a range of sizes; however, you cannot predict the shape or size of the gourd from the appearance of the seeds.

Ornamental gourd seeds

Luffa gourd seeds

ALL ABOUT SEEDS

One of the most important factors to insure a healthy and satisfactory gourd crop is the choice of seeds. Seeds contain not only the genetic code that will dictate the shape and size of the crop, but also the strength of the gourd shell and resistance to disease and pests. It is very important, then, to select the seeds with care before investing the time and energy into cultivation.

Gourds within a species, such as Lagenaria (hardshell) or Cucurbita pepo v. ovifera (ornamental) hybridize very easily. This means that it is often difficult to predict with certainty which exact shape and size of gourd will grow from a particular seed, unless it was specifically isolated or hand pollinated. Each gourd contains hundreds of seeds and each seed has been fertilized by a separate grain of pollen. Like individual children in a family, each gourd may have a very independent 'personality' or shape, varying from generation to generation.

People have often wondered if there is any way to predict shape or size of a gourd by examining the seed. Unfortunately, the answer is no, other than identifying species of gourd. Different genus and species have very individual seed characteristics (i.e. luffa gourds have small smooth black seeds and ornamental gourds have small smooth pale seeds). The seeds of the hardshell gourd are generally long (between 1/2 and 1 inch) flat and have slightly ruffled edges with 'horns' at one end. Within the hardshell variety, some of the seeds may be larger than others but this cannot be used to predict the size or shape of the gourd which it will produce.

If it is important to have gourds of a specific shape and size, get seeds from a reliable source. (See appendix of suppliers) Some horticulturalists raise single varieties of gourds specifically to maintain purity of strain. By hand pollinating each blossom, and then protecting it from any possible contamination by other well-intentioned insects, they are able to insure that the seeds will produce 'true' varieties. Usually these seeds come in small envelopes of approximately 10 seeds. It is impossible to predict germination rate, but 75% is generally accepted as the minimum to expect. Of course your own growing conditions can affect this. However, this count is a way to estimate the number of seeds to purchase and to plant.

Some commercial seed companies sell gourd seeds which are readily available in nurseries and garden centers, or through catalogs. They are broadly classified, such as 'birdhouse,' 'bushel,' 'dipper,' or even 'assorted.' In general, these seeds have a very satisfactory germination rate and are true to variety, within their broadly defined categories. (See appendix for listing.)

If you live in a cooler area, you may want to use seeds which have been treated to prevent fungal diseases. Many specialty gourd seed suppliers offer this option.

Many artists and craftspeople have found that they particularly like one gourd for its shape, shell quality, or some other characteristic. If you are working with a gourd you particularly like, remove the seeds from that gourd and save them to plant the next year. Two possible complications may affect the germination rate.

- If the gourd from which you collected the seeds was allowed to *freeze* while it was green or curing, the seeds will not germinate. Freezing kills fresh seeds. (Dry seeds can be stored in the freezer without affecting germination rate, however.)

- If the original gourd was fertilized by pollen from many other varieties of gourds, it is impossible to predict the genetic characteristics of the individual seeds it contains. Seeds saved from fruits that were crossed will exhibit hybrid variation when grown out. Successive generations will carry multiple genetic structures, and will produce a very interesting, but unpredictable, harvest.

Gourd seeds can be stored for up to three years if you take some simple precautions:

- First, the seed has to be very dry before it can be stored.

- Put the seeds in plastic packets, or even film canisters, clearly marking the shape, date and any other information you may need at planting time. Put these packets in a larger jar or container with a tight fitting lid, and store in a cool place. A cool, dry basement is ideal. Seeds may also be stored in the refrigerator, or even in the freezer, for up to several years.

- Because it is important to keep the seeds cool and <u>dry,</u> you may want to include a moisture desiccant in the container. A simple way to get a desiccant is to save it from a packaging that you may have received. Check the local pharmacy or drugstore for other possible sources. Or make a simple "moisture trap" by putting two tablespoons of powdered milk (from a freshly opened container) on a paper towel. Roll it up to make a small pouch and secure with tape.

If there is any concern about the viability of seeds, test for this prior to planting. There are many ways to make a seed "incubator." First, soak the seeds for up to 24 hours in warm water, and then use one of the methods described:

- Place at least 10 seeds between two paper towels that are kept damp and warm. After a few days, some of the seeds will begin to sprout. However, some seeds may take up to 7 to 10 days to germinate.

- Thoroughly soak several layers of newsprint. Place seeds between two damp paper towels, and then fold in the newsprint to create a sandwich. Place this in a plastic bag and store in a warm place. Check every 1-2 days for signs of sprouting.

- Seeds can be placed between damp paper towels, and then rolled between paper towels or terry towels. Stand the towel roll in a tall glass with water in the bottom. Store in a warm place for several days, checking frequently for signs of sprouting.

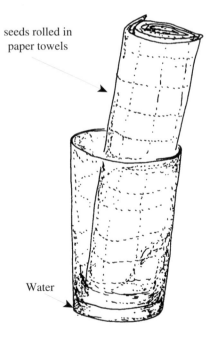

seeds rolled in paper towels

Water

Place seeds in a paper towel roll, and then stand it in a glass contianing a small amount of warm water.

Sometimes I have been surprised by seeds in the ground which sprout long after the other seeds in the hillock that were planted at the same time. Seeds often seem to have their own time-table for sprouting. However, especially if you are planting in an area with a short growing season, it is a good idea to sprout some seeds prior to putting them in potting soil if there is any question about seed viability. That way you are assured that the seeds you are planting will grow, and you won't have to start more seeds at a later date.

Seedlings often suffer 'transplant shock' when they are handled and moved about. They should be transplanted into peat pots as soon as there is any sign of germination. Once the seedlings have started growing in the peat pots, they should be kept moist, warm and under good lighting until the ground outside is warm enough for planting.

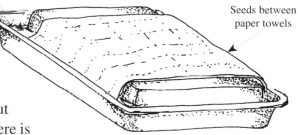

Water

Seeds between paper towels

Invert one flat pan inside a larger pan. Place seeds between two layers of paper towels, and lay across the smaller pan. Fill the bottom pan with water so that the edges of the paper towels are in the water.

PLANTING THE GARDEN

Gourds have a very long growing season, relative to many other vegetable crops. Ornamental gourds mature in the shortest time, as little as 100 days from planting. Hardshell gourds require at least 120-140 days to maturity, and luffa gourds require a minimum of 140 days of full sun.

Gourd plants are sensitive to temperature. Seeds and seedlings cannot be planted in the ground until well after the last frost in the spring. Ideally seeds or seedlings should not be put in the ground until the soil reaches 70 degrees. The first substantial frost of fall usually freezes or damages the vine and leaf cells. Gourds which are on the vines at this time that are not fully mature will rot instead of drying with a hard shell.

Unlike other fruits and many members of the Cucurbit family, gourds cannot be picked or harvested before they are fully mature, or they will rot. (See appendix or pp. 28-29 for planting zone guidelines.)

Snip shoulders off seeds with a fingernail clippers to promote faster germination.

Plant seeds in peat pots (or similar sized containers).

A plastic bag helps to retain moisure and maintain a warm environment.

Sprouts will begin to appear within 7 - 10 days

Once plants reach to four leaf stage, they are ready to be transplanted outdoors.

Starting the Plants Indoors

If you live in an area with a short growing season, it will be important to start the germination process indoors. Gourd seeds are encased with a very strong protective covering. Soak the seeds for up to 24 hours prior to planting to reduce the length of time till germination. (Do not soak longer than this, because seeds may rot!) Many growers also make small slits in the gourd covering to hasten the germination process (see illustration) Use a fingernail clipper to snip off the "shoulders" of the seed—a fast method that does not disturb the interior of the seed.

When starting seeds indoors, specifically designed peat pots of many different sizes are available in garden sections or nurseries. For gourd seeds, start with a larger size, at least 4 inches across, to give the rootlets plenty of room to grow without becoming rootbound or having roots exposed during the process of transplanting.

Any number of other similar sized containers that can be used instead of peat pots: cut off milk cartons, cottage cheese/yogurt containers, even small plastic baggies. Just be sure to provide drainage holes in the bottoms.

Fill the containers with soil mixtures specifically designed for seed sprouting (available in most nurseries or garden supply stores), place them on a tray and thoroughly moisten the soil. Plant the seeds approximately 1 inch deep, with 2 or 3 seeds per container. Cover the entire tray with plastic (sheet or bag) to keep the containers moist and warm. Check daily, and when the seeds begin to sprout, remove the covering. Once you notice growth, place the tray in full light, all the while keeping the containers moist and warm.

The first "leaves" to appear are not really leaves but cotyledons, smooth-edged and round in shape. The next leaves to appear are the real leaves of the plant, with irregular edges and five-lobed outline. Plants can be raised in containers until they reach the 4 "true leaf" stage, or approximately 4 – 6 weeks.

(Some growers will let plants stay in large peat pots much longer than this. However, it is safer to aim for a transplant date shortly after the plants reach the 4 leaf stage.)

NOTE: Many growers recommend using "grow lights" at this stage, since a sunny window probably will not provide enough light for seedlings to do well. (When seedlings grow next to a window using only natural light, they tend to get very spindly and leggy, and do not transplant well.) Grow lights promote leaf growth on shorter, healthier stalks.

There are many commercial "grow-lights" on the market today, but it is not difficult to construct your own. The minimum materials needed is one 24" long fluorescent light fixture that holds two bulbs. To accommodate more seedlings use a 48" long frame. Mount it under a cabinet with two chains so that the height above the plants can be adjusted as the plants grow. Keep the light 2" above the tops of the plants, and allow 14 hours of light a day.

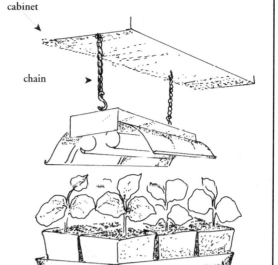

cabinet

chain

To make your own "grow light" suspend a fluorescent light under a cabinet. Chains and hooks will allow you to raise the light as the plants grow.

One week prior to planting outdoors, the seedlings should be "hardened off." To accomplish this, place the trays of seedlings outdoors next to a south facing wall, or in a similar warm, protected spot for a few hours in the warm part of the day. You can follow a schedule of gradually increasing the hours that the plant is outdoors until the plant is able to remain outside for 24 hours. This process should take about one week. When the frost is safely past and the days are consistently 70 degrees, transplant the seedlings to prepared beds, disturbing the roots as little as possible. There are several ways to further protect the seedlings, both from weather, bugs or inquisitive skunks or nighttime explorers:

• Pre-heat the ground by covering it with black plastic sheeting prior to planting. Depending on the weather and growing conditions, this can be laid down up to several weeks prior to planting. When it is time to put the seedlings in the soil, either move the plastic over to reveal a row of soil between the sheeting, or simply cut X's where the seedlings will be planted.

Plastic bottles make good plant protection for new seedlings.

Drill holes in the side of an old tire, and then use it as a protective barrier for seedlings

Cut bottom from a cardboard box, and place around seedlings. Fold flaps over to protect plants from sudden frost.

Put Wall-o-Water around seedlings to protect against sudden frosts. Remove when plants begin to vine.

A rigid plastic sheet that can be draped over a hillock or row. Remove once seedlings begin to vine.

- Cover each seedling with a strawberry basket as soon as it is planted.

- Cut the bottom and top off a milk carton or plastic soft-drink bottle. Push into the ground enough to make a protective surround, and remove when the plants are larger and stronger.

- Make the hillocks in old tires which are sunk into the ground. The tires can be covered with sheet plastic if the weather turns cold, and the black rubber collects and retains heat.

- Cut the bottom off a shallow cardboard box, and set it over the hillock with the top flaps open. If the weather turns cold, simply fold the top flaps over to protect the seedlings.

- Purchase a protective device made of a circle of connected plastic pillows (Wall o' Water). Filled with water, they absorb heat during the day and release it at night, shielding the young plants from cold nights.

- Cover your seedlings with row covers, or "floating" horticultural mesh. This can be braced above the seedlings by blocks, bent metal or PVC frames.

Whatever physical protection provided, many growers recommend dusting the groundsurrounding the plants with Sevin or other insecticide. New seedlings are especially delectable to garden pests.

Seedlings that are transplanted have a certain setback period, as the plant readjusts to its new environment. Because the plant was started indoor six weeks prior to planting may not mean that you have added six weeks to the growing season—the plant may take up to two weeks to recover from transplant shock.

Planting seeds directly into the ground

In an area with a long growing season, you can plant the seeds directly in the ground. The germination time may take up to two weeks unless the seeds have been trimmed and soaked (see above description). Be sure to keep the soil moist after planting.

Prepare the soil up to one month prior to planting.

I have planted seeds in the ground, and then covered the rows with black "plastic" or horticultural mesh. A soaker hose allows for watering the ground as needed, but the black cover reduces the evaporation and keeps the soil very warm. As soon as the seedlings begin to appear, move the black plastic over so that the ground is covered except for the thin row where the seedlings appear. This maintains an overall warm soil temperature, allows for much more efficient watering, and greatly reduces the need for weeding! (In some areas of the country, it is possible for the black plastic to become too hot during the middle of summer. In this situation, cover the plastic with straw or mulch to moderate the temperature and prevent it from burning the plants.)

The first leaves to appear are not "true leaves" but smooth-edged cotyledons. The true leaves appear soon after.

The gourd plant is a slow starter. Initially, it takes a relatively long time for the seeds to finally germinate and begin to peek out of the ground. By three-four weeks, the plants will be at the four-six "true leaf" stage, and you will notice that some seedlings are hardier and stronger than others. This may be a bit confusing if you have planted 'assorted gourd seeds,' since some varieties may be slower starters or may be smaller plants initially. Look for the plants that appear "stronger," rather than "larger." As the plants begin to grow, thin them out to two plants per hillock. If you have planted the gourds in a row, allow one plant every four feet.

If you are planting gourds in a container, gauge the size of the gourd and the container. As a general rule, one or two plants per large pot is sufficient. (I have experimented by growing 3 plants per oak barrel, and found that the vines are crowded but still produced an abundant crop of mini-ornamentals.)

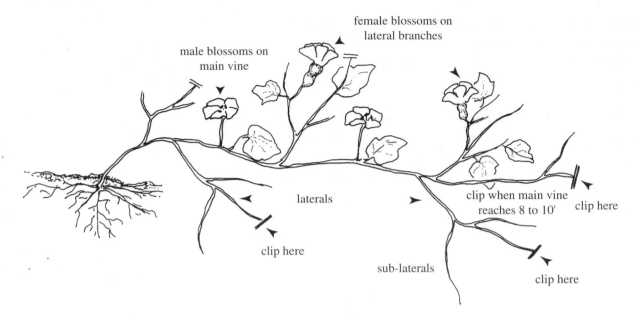

female blossoms on
lateral branches

male blossoms on
main vine

laterals

clip when main vine
reaches 8 to 10'

clip here

clip here

sub-laterals

clip here

Trimming the Vines

The young plant produces 6-8 more leaves and then seems almost to rest, growing very little. Actually, the action is happening below ground, as the root system becomes established. After seven-ten days of apparent inactivity, the young plant will begin to vine. When the primary vine reaches a length of 10 feet, nip off the tip to force growth in the lateral branches. There are two reasons for this: The most important is that the gourd plant is monoecious; that is—male and female blossoms grow on the same vine. However, the male blossoms grow on the primary vine, and the female blossoms appear on the lateral and sublateral branches. The female blossoms are the ones which produce the fruit, or the gourds. Since the goal is to have more gourds, you need to encourage maximum growth in the lateral branches.

Another reason for nipping the primary vine is that the gourds are very vigorous growers, and untrimmed vines can reach up to 100 feet in length. Most gardens simply don't have room for this kind of tenant! Untrimmed vines frequently climb up tall trees and smother surrounding bushes and plants. Trimming back the vines will promote the growth of sub-lateral branches and fruits, which will result in a much more abundant gourd crop as well as create a more manageable plant.

When growing gourds on a trellis, follow a slightly different procedure in the trimming. Initially remove the laterals from the main vine until the vine reaches the top of the trellis. At that point, nip the main vine to promote the growth of laterals along the top surface of the trellis.

As the vines continue to grow, they will develop tendrils which will grasp the supports quite securely. The tendrils usually occur at the axil where the female blossom emerges from the vine, next to the stem of the gourd. As they first appear they are straight and branched, but as soon as they find something to grasp they quickly encircle it, first in one direction and then the other. Although the tendrils appear delicate, they are extremely strong.

Continue to water the gourd plants as they begin to 'vine.' They should have at least weekly deep watering. Watch the leaves to determine if they need more. Different climates, heat, humidity and rainfall create very different needs in each garden. Gourds grown in pots will need more frequent watering —at least three times weekly or even daily.

If the soil was well fertilized prior to planting, the plants will not need additional fertilizer for 6-8 weeks. After that time, as the vines are beginning to extend, add a well balanced fertilizer that contains nitrogen. The nitrogen promotes strong leaf production and vigorous vine growth; however, nitrogen tends to inhibit flower and fruit set. Therefore, once the blossoms begin to appear and the baby gourds are beginning to set, any fertilizer should be low in nitrogen and higher in potash and potassium (0-10-10).

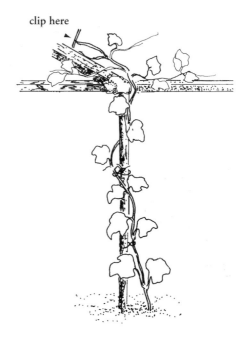

clip here

When training vines on a trellis, clip the laterals off the vine until the main vine reaches the top of the trellis. Then clip the main vine to allow the laterals to cover the trellis cover.

New seedlings and young vines are very susceptible to garden pests. (See pp 77.) Particularly if you are planting gourds in an area where gourds were grown in past seasons, some insects such as the cucumber beetle may have wintered in the soil and be waiting for a fresh morsel to devour. At the first sign of insects, dust the area around the plants with an insecticide. Be careful not to get too much dust on the plant itself, since young leaves may be burned or stunted by the chemicals. For alternatives to insecticides or chemicals, see section on pest control. Many natural remedies are available for the gourd gardener which may be very effective if used at the very outset before pests and diseases have a chance to become big problems.

Because gourds have a shallow root system, it is important not to disturb the soil once the plants begin to vine. However, weeds may steal nutrients and moisture which the gourd plants need, so keep the area around the gourds clear by hand or hoe. New products on the market include a garden mesh or plastic sheeting which is specifically designed to keep weed growth down and still allow moisture to penetrate to the soil. You can also surround the plants with straw or similar garden mulch. This inhibits weed growth as well as provides a good surface for the gourds to sit on as they mature.

POLLINATION

After the vine has reached a length of approximately 10 feet or has been growing six weeks, it produces many beautiful large delicate blossoms. The blossoms are cup shaped with five large petals, up to 3 inches in diameter. The ornamental gourds (Cucurbita pepo var. ovifera) have a golden yellow blossom, the luffa gourds have a delicate yellow and slightly smaller blossom, and the Lagenaria hardshell gourd has a beautiful larger white blossom.

Both the luffa and ornamental gourd plants bloom during the day, but the hardshell gourds bloom at night. The blossoms of all of the vines last only 24 hours.

On close examination you will notice that the first blossoms to appear come from the primary stem of the vine, standing upright on long delicate stalks. These are the male blossoms. The delicate blossoms last only one day(or night) and then fade and drop off the stalks. However, they are continually replaced during the growing season with an abundance of new flowers. The female blossoms begin to appear several days later on the lateral and sub-lateral branches of the vine. They are distinctive from the male blossoms in many important ways:

- Directly under the petals there is a large swelling. This is the "pepo," or the berry that will become the gourd fruit if the blossom is fertilized. This tiny gourd is a miniature of the shape of the adult gourd if the blossom is sufficiently fertilized.

- The blossoms are on a relatively short stalk, nestled in axils of the leaves—that is, where the leaf stem joins the plant.

- Comparing the male and female blossoms, the differences between the two become immediately apparent. The male blossom has five sepals, five petals which are fused, and five stamen topped by the anthers that carry the pollen. The female blossom, besides having the swollen "ovary" under the petals, has 3 two-lobed stigma which curl and extend up from the center of the blossom.

White blossoms of the hardshell gourds float above a sea of green leaves.

Golden yellow male and female blossoms cover the ornamental gourd vines.

A petal has been removed to show the stigma and the pepo, or baby gourd, below the blossom.

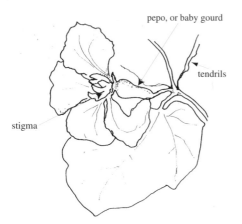

Female Blossoms

Natural Pollination

Fertilization occurs when the pollen from the male flower is placed on the stigma of the female flowers. The plants which bloom in the daytime (luffa and ornamental gourds) are pollinated by honeybees and other daytime flying insects.

The white blossomed hardshells, however, bloom only at night, and there is real question as to the pollinizing agent. Many people believe that it is a night flying moth. In some southern arid areas, a night flying bee may serve this purpose. In *The Gourd Book*, Dr. Charles Heiser suggests that the only beneficial purpose of the cucumber beetle may be to pollinize the gourd plant!

At any rate, pollination will occur naturally in an environment that has an abundance of flying insects. More and more, however, insecticides, pesticides and other naturally occurring blights have greatly reduced the number of these beneficial insects. Well intentioned gardeners who try to protect their plants from pests and diseases may inadvertently also be ridding their garden of the desirable insects as well. If you have been spraying the plants, stop once the blossoms appear. Bees will avoid blossoms that have been sprayed.

Hand Pollination

To insure that more blossoms are fertilized, thus increasing the gourd crop, you can assist nature by hand pollinating the plants. This is a very easy process, once you learn to identify the male and female blossoms.

There are two ways to hand pollinate:

- With a small soft bristled paint brush, gently collect the pollen from the male blossom and lightly dust it on the stigma of the female blossom. Because every seed in the future gourd plant is fertilized by a separate grain of pollen, dust the female blossom with pollen from more than one male blossom.

- A second method is slightly more direct. Pick the male blossom and gently peel away the petals to expose the anthers and pollen. Gently dust this over the female blossom, giving it a gentle flick with your finger.

If the female fruit has been adequately fertilized it will soon swell. The blossom will shrivel and drop off, and the gourd will quickly begin to grow. If the blossom is not fertilized, it will drop off, and eventually the swollen stem, or fruit bud, will also become brown and drop off.

Occasionally a baby gourd will begin to grow, and within a week will turn brown and drop off. This is probably because the female blossom received some pollen, but not enough to develop sufficient seeds to make full gourd growth viable. Mother Nature will not waste vine resources on a gourd that is not going to produce seeds for the future.

Ornamental gourds and luffa gourds can be hand pollinated any time during the day, since they are day-blooming plants. Hardshell gourds that blossom at night should be hand pollinated in the evening shortly after the female blossoms fully open. On long summer evenings, this is a delightful task in the fading light of dusk. By the next day the blossoms will begin to fade, wither and drop off.

When pollinating the blossoms, you may want to mark them to keep track of their growth. This can be done in several ways:

• Put a string or yarn tie around the vine next to the female blossom.

• Slip a small plastic ring over the blossom which will remain on the stem until the gourd is harvested.

• Spray paint a mark on a nearby leaf.

If the goal is a crop of unusual and interesting gourds, it doesn't matter which vine the male and female blossoms are from. (However, while this act of pollination is necessary to fertilize and insure that a gourd will emerge from the female blossom, the pollen is actually fertilizing the seed within the gourd. That means that the shape of the gourd that appears this year will not be affected by the pollen of the male blossom; the seed *within* that gourd will carry the new genetic code to influence the next year's fruit.)

The pollen is clearly visible on the anthers of the male blossom.

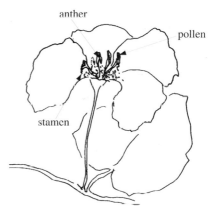

Male Blossoms

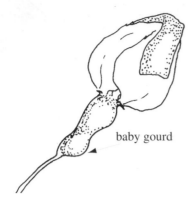

baby gourd

Tape petals together once the female blossom has been pollinated, (above).

Or, cover the female blossom with a bag to protect it from further pollination.
Remove when the blossom falls off (below).

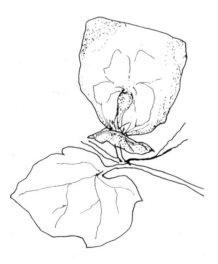

If the goal is to produce a crop of gourds that will provide seeds "true to type" or having the same characteristics as the parent (or grandparent) gourds, fertilize the female blossom with pollen from male blossoms on the same vine or from a vine of the same gourd variety, and then immediately protect it from other insects. You can do this easily by folding the petals up over the stigma and taping them together carefully. Some growers protect the female blossoms with small bags which they tape around the stem, completely covering the blossom and tiny pepo. Leave this protective covering on the plant for a couple of days, and then remove when the blossom is ready to drop off.

Many growers are trying to separate and preserve gourds that are "true to type" or variety. This is very difficult since the varieties have hybridized so completely over the years. At this time no one is completely sure what is a true, or original, variety. Also, it takes many generations of care to isolate a completely reliable and "true to type" seed. These growers always hand pollinate the gourds, using male and female blossoms from the same vine or the same variety but a separate vine to maintain some genetic diversity. They take special care to isolate the female blossom so it will not be pollinated accidentally by an insect. Once the blossom is pollinated, it is "bagged" or sealed and marked with identification colors or numbers. Fields where only one type of gourd is being raised for seed are kept isolated from other gourd plants to prevent any accidental cross pollination. A distance of one or more miles between fields is recommended.

For a more complete discussion of this process, please refer to the book *Seed to Seed*, listed in the bibliography. Author Suzanne Ashworth has been working for many years with the Seed Saver organization to isolate and preserve many heritage seeds, including many types of gourds.

Cross Pollination with Other Crops in the Garden

Much confusion exist about cross-pollination among members of the large Cucurbitaceae family. Remember that cross-pollination can occur only between varieties within a species. Since the hardshell gourd (Lagenaria siceraria) has only one member in the genus in North America, it will not cross-pollinate with any other Cucurbits. However, the varieties within the species readily cross with each other, hence creating the wide variety of unpredictable shapes and sizes seen in most gourd gardens.

Luffas also are in a separate genus, and do not cross pollinate with other cucurbits. However, the ornamental gourds belong to a genus and species that has a very great many members (Cucurbita pepo). This includes many of the summer and winter squash and even some pumpkins. Thus the ornamental gourd has the potential to cross-pollinate with many other plants in the garden. While the results will not be apparent the first year, this will affect the seeds in those fruits, and the second year some unusual colors, textures and flavors in the volunteer squash may appear in the garden area.

FRUIT SET
Thinning Excess Fruit

Soon after the blossoms begin to appear on the gourd plants, you will notice the first gourds set.

Baby gourds are attractive not only to you, but to a host of insects and diseases. Take special precaution at this time to protect the crop. If the gourds are growing on the ground, rest the small gourds on a support to keep them off the moist dirt and away from snails, slugs and other potential bacteria the ground may harbor. Sturdy paper plates are very useful for this purpose; also plastic bags, cardboard, wooden pallets or a bed of straw have been used successfully. Sprinkle a light dusting of insecticide on the support if there are slugs, snails or other crawling insects. You may want to lightly dust or wipe the young gourd with a mild insecticide to protect against flying insects.

Baby gourds can be placed on a piece of wood or even a paper plate.

*A design was scraped in
an ornamental gourd
by Bill Kupka.*

The gourds that are first to set usually attain the largest sizes and also the thickest shells. If you live in an area with a short growing season, or if you want larger but fewer gourds, thin the crop of baby gourds to encourage all the energy and nutrients of the vine to be directed to a few gourds. Male and female blossoms will continue to appear on the vines throughout the growing season, but late set gourds will usually not have time to mature before the first frost kills the vines. Some growers just remove all of the female blossoms that appear once they have determined which gourds they want to nurture for the remainder of the growing season.

Shaping/Training Gourds as They Grow

Most gourd growers are fascinated by the plant itself, and by the wondrous variety of shapes, sizes and colors of fruit that Nature creates. The gourd is certainly one of the most original and varied members of the plant kingdom. Because the plant is so ancient and so adaptable, it is the only plant that is found naturally occurring (that is, not introduced by man) throughout the temperate and tropical areas of the globe. Humans found such diverse uses for the plant, that either through natural mutation and hybridization, or by accidental or intentional cultivation, shapes and sizes of gourds can be grown to meet almost any imaginable use. Given the variety that naturally occurs, most of the time the gourds are allowed to grow and become whatever the seed dictates. Even on a single vine, two gourds are not identical.

Some growers, however, are intrigued by the possibilities of what can be done to modify the original gourd shape. Many different techniques can be explored if you want to experiment with gourds while they are still growing.

Scraping Designs in the Skin

While the gourd is growing, take a sharp pointed instrument (a nut pick, dental pick, awl, etc.) and carefully scratch or scrape a design in the thin epidermis of the gourd. Be careful not to puncture or penetrate through the gourd wall because that will cause the fruit to rot. As the gourd continues to grow, a scab or ridge will grow over the scraped area as a natural protection against invasion by insects or disease.

Any kind of bruise, scratch or scrape such as this will make the gourd vulnerable to disease or bacteria. To provide some protection, carefully wipe the shell with a very mild solution of insecticide, or a light dusting of Sevin. Examine the shell frequently to make sure that it continues to grow healthy. Occasionally reapply the insecticide to make sure the shell remains protected. Once the gourd is harvested, the scratched design will appear as a permanent raised outline on the shell.

Molding Your Gourd

Often gourds will naturally grow in such a way as to have a flattened spot on the side or bottom where they rested on the ground. Occasionally small gourds will lodge in spaces that restrict their growth, such as between slats of a fence, draped over or around a pole, caught in a wire fence. To expand on this idea, many growers through the years have created molds to influence the shape of the gourd shell.

Put a gourd between two flat boards, and fasten them together with bolts or wire.

- The simplest mold consists of two flat boards which can be secured to either side of a gourd and then tied or wired together. As the gourd grows, it will become flattened and perhaps elongated.

- While a gourd is small, a ring or series of rings can be slipped over the gourd and positioned in such a way as to create multiple lumps, or bubbles, in a gourd.

- While many bottle gourds naturally have a narrow waist between two bulbs, in prehistoric times people in some cultures used to constrict the waist of all gourds with a wrap of twine to create a handy way to tie gourds on to fish floats, or to fasten handles onto water carriers. Rings of wood or metal can be slipped onto bottle gourds to create not only useful but decorative elements as well.

This gourd was grown in a plastic vegetable mold.

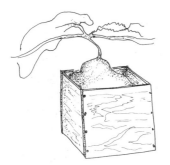

Put a small hardshelled gourd in a wooden box to create a square gourd.

To grow a gourd inside a bottle, be sure to tie the bottle to the trellis for support.

- A netting can be tied around a small gourd which will constrict future growth. While in other times this technique was used to shape gourds that could carry water and other valuables securely, this method is used today for purely decorative purposes.

- Gourds can be grown in molds to create a wide variety of shapes and surface designs:

    A gourd grown in a wood box will become a cube.

    Insert a small gourd in an unusual shaped bottle to create one-of-a kind designs. (The bottle must be broken when the gourd is finally mature so that the moisture can evaporate as the gourd cures.)

    Molds can be purchased that are designed for pouring clay or plastic items or even growing vegetables. (Vegiforms, 2 Burton Woods Lane, Cincinnati, Ohio 45229) Many of these can be adapted to gourds as well.

- The Chinese were masters of molded gourds which they created to house their prized fighting crickets. Many of the cricket houses were very plain, but some had elaborate designs which embellished the exterior of the shell. The molds which they made were created in several sections (a minimum of 4). The insides of the mold were either left smooth or were carved with elaborate pictures, designs, or inscriptions. This was then tied around a small gourd, which expanded to fill the interior of the mold. When the gourd was mature, the mold was either untied or cracked off the gourd to allow the gourd to dry and completely cure.

If you grow a gourd in a mold, there are some important precautions to follow:

- If the mold and gourd are on the ground, the mold should have plenty of cracks or holes for drainage and ventilation. Standing water will cause a small gourd to rot inside a mold.

• It is preferable for a gourd in a mold to be suspended on a trellis. The mold should be supported on the trellis independent of the gourd vine and should be positioned in such a way that the opening will not get rain or moisture in it. Check frequently to make sure the gourd is not developing a fungus or soft spots under the mold. Young gourds are delicate and sensitive.

Some growers are challenged to create unusual shapes with gourds that have long and narrow portions, such as dippers, snake gourds, clubs and marankas. If these gourds are grown on the ground, they will naturally assume unusual shapes as the gourd tries to push and turn to assume its natural length. When grown on a trellis and allowed to hang freely, these gourds will be long and straight. Baby gourds can be trained into special shapes if they are handled with extreme care. It is best to work in the heat of the day when the baby gourd shell is soft and flexible. If they are to be wrapped around a frame, pole, rope, or other stationary object, then that object should be firmly secured to the trellis next to the gourd. The gourd can then be twisted gently to the desired shape and tied with a soft cotton rag.

If the shape requires tight corners and twists, such as when tying a knot in the neck of a long-handled dipper, it may take several days to accomplish the full pattern. Begin the training when the baby dipper gourd is approximately 6 inches long or less, and always work in the heat of the day, when the gourd is most flexible. Select a dipper that has a long stem, which can be used during the initial training process. Carefully bend the bulb up to form a partial circle, and tie securely with a soft cotton rag. In subsequent days gradually bring the bulb up and around the stem, until the bulb just barely extends through the loop. Don't try to cinch the knot, as the weight of the bulb as the gourd grows will complete the knot. Most of all, be patient. Small gourds have a breaking point, and be prepared to snap several as you try to create your perfect one-of-a-kind creation.

This Chinese cricket cage was grown in a mold to create elaborate designs on the shell

This gourd (left) was grown in a netting to produce a regular bulbous pattern.

These two gourds (right and below) were grown in a variety of molds by Jim Story

For most other training patterns, support both the gourd and the form so that undue weight or stress on the vine won't accidentally snap off the gourd.

Once you have trained the young gourd into the position or shape you want, leave it alone to mature on the vine. Once the stem is brown and the vine dies back, cut down the gourd (leaving 1-2 inches of stem), remove the mold or training forms, and allow the gourd to dry and cure naturally.

Photo: by Jim Story.

Begin to train a dipper gourd into a knot when it is very small.

Trained by Jim Story.

This extra-long handled dipper was also trained by Jim Story.

HARVESTING YOUR CROP

I was recently approached by a young couple very interested in gourd craft, especially for making musical instruments. For several years they had been trying to grow their own gourds to make rattles, and had no luck—the entire crop rotted every year! I suggested they may have an infestation of bugs, fungus or some other disease that infected their crop, but they assured me the plants and fruits were healthy and strong. The gourds looked perfect until they "picked them when they reached the right size for the rattles, and right away the gourds just rotted away!"

Many people (like this young couple) don't realize that you cannot pick a gourd "when it reaches the right size." Gourds must remain on the vine until they are fully mature. The gourd continues to grow and receive nourishment from the vine until at last the stem begins to turn brown and the tendrils near the gourd are completely dry.

Ornamental gourds

With small ornamental gourds, this point of maturity is reached sooner than with other gourds. In late summer, many small ornamentals will have dry stems, and can be safely cut from the vine. To preserve the bright colors of the ornamental gourds, store them in a cool place out of the direct sunlight—the sun will tend to fade the colors, or turn the white skins into an ivory or light tan shade. The vine will continue to grow and produce more gay and fanciful gourds to complement your fall decor.

The *Cucurbita pepos* or ornamental gourds can be harvested any time once their stem turns brown and the tendrils next to them are dry. When ornamentals are mature, they create a fascinating display of unpredictable colors, shapes and textures. They are truly one of Nature's show-offs, and are one of the important harbingers of fall as we begin to see them in grocery store displays. Unfortunately, much of their beauty is only skin deep—the epidermal layer which covers the gourd shell carries the color. When the gourd loses its moisture and the epidermis eventually molds away, the gourd becomes a light tan woody shell. If they are stored in a cool area, and nothing is done to them, they will eventually dry and become thin shelled woody permanent gourds, a miniature of their hardshell cousins.

People have tried to preserve the bright colors in many ways. Store owners frequently will varnish or shellac the gourds. This makes the exterior even more brilliant, but unfortunately clogs the pores of the gourd, so instead of evaporating, the moisture inside rots the shell and it must be discarded after several weeks or months.

The gourds can be polished with a light furniture wax or salad oil. This will highlight the colors, but does not clog the pores, so eventually the gourd will be able to dry naturally.

Several recipes have been suggested to preserve the colors of the ornamental gourds indefinitely. The one method recommended by the American Gourd Society is as follows:

1. Prepare the following solution:

    1 cup 20 Mule Team Borax

    3 cups hot water

    Dissolve all the borax if possible in
    cool to luke warm water.

2. Dip gourds in a pan of boiling water briefly. Then place them in the borax solution and leave for 15 minutes. Set a timer to be exact. It is possible to do many at one time.

3. Remove the gourds but do not rinse. Place in a wire basket and hang where they will get plenty of air circulation. After several weeks check to see if they are completely dried. Wash in warm water and then wax with a paste wax. (I have tried this and several other methods to dry ornamentals while leaving the colorful epidermal layer intact, without much success. Usually the colors fade substantially.)

Luffas

Luffa gourds should be left on the vine until the stem is completely dry and the gourd itself is turning brown at both ends. As you squeeze the gourd, you can feel that the shell has separated from the fibrous interior mass, and hear the seeds rattle inside. Luffas can be harvested individually as they reach this stage, or they can be left on the vine until they, and the vine, are completely dry.

Luffas can be peeled at any point once they are harvested. The *Luffa acutangula*, the ridged variety, is usually grown for food, and is very difficult to peel. With the more familiar varieties, the shell is thin and can be split open with a fingernail. Stringy fibers grow lengthwise in the shell, and can be pulled down to further split the shell apart. The skin can then be peeled away from the spongy interior quite easily. If the shell has been allowed to dry completely, you can peel it away bit by bit—like peeling a hard boiled egg. You can also soak the luffa gourd for 15 minutes to 2 hours. This softens the shell and the pieces will peel away with relative ease.

The interior sponge will be rather slippery and shiny if the shell is still partially green when it is peeled away. Otherwise, it is a light tan dry sponge filled with 40 to 80 seeds. Soak the sponge in warm water for a short while and then hang up to dry. Many people like to put a small amount of bleach in the water to remove any discoloration and produce a sparkling luffa sponge. Save the seeds of the particularly good sponges for next year's crop.

These Luffas are beginning to dry.

Hardshell gourds

The hardshell gourds generally require a longer time to reach full maturity, and most likely will not be completely mature until fall. A crop of mature hardshell gourds (Lagenarias) presents an unexpected variety of colors from ivory to dark green, with interesting patterns and textures in addition to the shapes. Like their more colorful little cousins the ornamental gourds, the beautiful colors and patterns are only skin deep. There is no known way to preserve these successfully.

Some may mature earlier than others on the vine; if the stems and tendrils next to the gourd are completely brown, the gourd can be carefully removed from the vine. Usually, however, plan to keep the gourds on the vine and in the fields until after the first good frost, or until the vine itself begins to die.

In addition to the brown stem, the gourd itself will feel firm to pressure at the base of the stem, and may be beginning to lose color and turn an ivory shade. It may even start to lose some weight—that is, some of the moisture may be starting to evaporate. When harvesting the gourds, be sure to cut the stems several inches above the gourd. The stem is considered to be an integral part of the gourd, and many artists and craftsmen integrate the stem in their design. (Any gourds entered for competition or show must include 2-5 inches of stem along with the gourd.)

Gourds that were late to set on the vines may not ever have a chance to completely mature. They may be cut off the vine and even stored, but will probably begin to soften and rot, as opposed to harden with a woody shell.

Leaving the gourds in the field

Many growers just leave the gourd crop in the fields over the winter, allowing them to dry where they grew. It is perfectly safe to leave them out in winter weather. The alternate freezing and thawing will not harm the shell. However, the freezing weather will affect the seeds. Seeds which have frozen while green will not germinate. Therefore, if there is a gourd that you want to save for seed, be sure to bring it into a protected area. The disadvantage to leaving the gourds in the field is that the vines will also be allowed to cover the ground. If they had any disease or pest infestation, this will be transferred to the soil and will be much more difficult to eradicate in the following growing season.

Gourds can be left in the garden long after the vine has died. (Garden of Helen Bos)

Bringing the gourds in from the garden

Many people prefer to bring the gourds to a central place as they dry and cure. Gourds that have been grown on trellises should definitely be harvested and moved. Dry vines may split and probably will not support heavy mature gourds, particularly in winter weather.

It is important to store the gourds in an area where they will have good ventilation, since this is the period of time when the moisture inside the gourds will be evaporating. Ideally, gourds should be spread out on a surface such as wire mesh, so that they have complete ventilation. You can also wash them in a weak solution of disinfectant or laundry soap with some bleach added. This will clean away any soil and possible bacteria or pests which were present in the field and which could affect the gourds as they dry.

Although it is not critical, many growers like to rotate or turn the gourds as they dry, thus preventing soft spots from developing where the moisture gathers in the interior.

The surface of some gourds may have been damaged by garden pests, but this may not eliminate their use by a crafter. As long as the hole or scar does not penetrate the shell completely, it can be repaired once the shell is completely dried and cleaned. Frequently artists are able to use defects as a source of inspiration for their designs!

The most important thing to remember, however, is this: Do not attempt to drain the water out by puncturing holes in the gourd shell. The water does drain, but bacteria will enter the hole, and the entire gourd may rot. You can safely drain the water out by drilling a large hole, minimum diameter 1" (see pp. 73), and frequently this method is used to prevent a very large gourd from rotting from the inside. The secret is the size of the hole: the smaller the hole, the greater the potential for trouble.

It is safe to harvest gourds if the stem and tendrils next to the gourd are brown. (Garden of Helen Bos)

The time it takes for gourds to completely dry out is difficult to predict, since it is as individual as the gourds themselves. It may take as little as several weeks or as long as a year, depending on the type of gourd, size, and thickness of the shell.

There are many factors which will affect evaporation:

Color fades from the epidermis as the gourd begins to dry.

- **Temperature:** gourds stored in a warmer and drier climate will dry somewhat faster than gourds in other climates. However, do not try to 'force dry' your gourds by putting them in a heated room (such as a furnace room). Gourds that are forced to dry quickly often develop cracks or shrivel and rot. The shells seem to have their own timetable for curing, and forcing this to a shorter time is rarely successful.

- **Shell thickness:** some gourds have thinner shells than others, or the shells are slightly more porous, allowing the water to evaporate more quickly. This may be somewhat misleading, however, because the thin-shelled ornamental gourds do not necessarily dry more rapidly than their thick-shelled cousins. Many remain much like they were when freshly harvested throughout the following spring and summer.

- **Size:** frequently smaller gourds will dry more quickly than the larger gourds, simply because they contain less water. Very large bushel gourds may take as long as a year to dry completely.

- **Shape**: Protrusions and extensions (such as dipper handles) may dry more quickly than the bulb of a gourd. Dipper handles usually begin to dry from the stem end; bottle gourds and lump-in-neck gourds have dry tops while the bases remain filled with water.

The first changes that occur are a fading of color in the epidermis. The bright greens or spotted and mottled coats gradually become a dull pale green or ivory. Eventually the epidermis will begin to discolor and show signs of rotting or mildew.

Inexperienced gourd growers often become alarmed and feel that their precious crop is beginning to decay. As long as the gourd shell feels firm to the touch, the curing process is proceeding naturally. (If a gourd does begin to collapse or show soft spots, get rid of it immediately so that it does not affect the rest of the crop.)

Some craftspeople prefer to wipe off the mold that forms with a mild disinfectant. The gourd continues to dry but will not develop the mottled appearance that is so often found on the gourd shells.

If left alone, gourds that dry indoors or in a warmer environment tend to develop a very fuzzy mold or "bloom" over the surface of the shell. This does not hurt the shell, and only serves to create interesting patterns or "mosaics" on the gourd shell.

Anywhere from 3-6 months after harvest (or up to a year for larger varieties) the gourds will feel lighter when you pick them up and the seeds may rattle. Under the patchy dried epidermis the gourd shell is a woody color, tan to light brown, and feels smooth and firm. Unfortunately, all of these clues are relative, and it is difficult to give a precise description of just when the gourd is truly dried out.

Some gourds have a very thick dense shell and will always feel heavier relative to other gourds of their same size and shape. Also, sometimes the seeds dry encased in a ball or lump of the inner pulp. This may sound like a thud instead of a rattle when the gourd is shaken. Or the pulp and seeds may have dried adhered to the inner wall so that there is no sound when the gourd is shaken.

The color of the exterior may range anywhere from creamy tan to reddish brown, with many shades and variegated patterns in between. Experience touching, feeling and holding gourds will help you become a judge of when the gourd is dry and ready for your future plans.

The epidermis molds and flakes off when the gourd is dry.

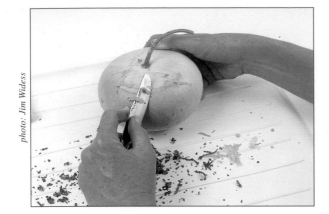

photo: Jim Widess

HASTENING THE CURING PROCESS

If you are impatient and do not want to wait the many months necessary to allow a gourd to dry naturally, you are not alone. Through the ages people have been anxious to use the vessels of their gardens without waiting for Mother Nature to finish the job. There are many different things you can do to hasten the process.

Scrape off the green epidermis.

The epidermis is a thin waxy coating over the gourd, which protects the shell as it is growing and even after it is mature and has been harvested. Until the shell dries to become a strong woody surface, the epidermis protects it from disease, fungus and numerous pests that may try to invade the shell

Use a dull kitchen knife or putty knife to scrape the epidermis off. Be careful not to scratch or mar the gourd shell. Once the epidermis is completely scraped off, use a copper kitchen scrub pad to remove any remaining bits of epidermis. The exposed gourd shell must then be watched carefully and wiped at least weekly with a mild solution of water and household detergent to keep mold from building up. The gourd will dry much more rapidly than those with the epidermis intact, and will have a clear ivory to slightly tan shell.

Occasionally gourds that have been scraped in this manner will shrivel and crack during the drying process. Select only fully mature gourds to scrape while they are still green.

The epidermis can be scraped off a gourd at any time after it has been harvested. By waiting for 1-2 months until the water has begun to evaporate from the interior, the epidermis will naturally be softening and separating from the shell. It is much easier to remove at this point, and the shell will still remain clear and free of the mottling on gourds that dry naturally.

You can cut the top off a gourd while it is still green.

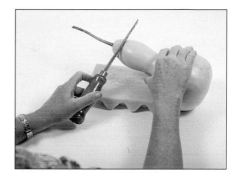

Drill a hole or cut the tip off a green gourd —

Use a saw, carving knife or drill and make a hole that is at least 1 inch in diameter, preferably larger.

The interior pulp will resemble that of acorn or winter squash—a fibrous mass with lots of seeds. Carefully scrape out as much as possible, and then fill the shell with water. Let it stand for a week or more, until the remaining pulp becomes a gelatinous muck, making it easy to remove. Then carefully scrape out the interior.

Once the shell is empty, you can scrape off the epidermis, and then let it completely dry and harden. Watch it carefully to remove any mold that may start to form. Occasionally gourds that are worked in this manner collapse slightly, particularly around the cut edge. Therefore, make the initial cut slightly above where you will want the final edge of the gourd container so that it can be trimmed down once it is completely dry.

— then fill with water to rot out the inside.

This method was used by many prehistoric tribes in Africa, New Zealand, and the Americas, but has not been tried by contemporary artists and craftspeople. One problem is that many people have not yet decided how they want to cut or use a gourd while it is still green. If you want to make a simple bowl or large open container, this is a very effective method to use.

This is also the technique used by Japanese to clean out sennari gourds that became containers for sake or water. While green, a hole is drilled in the tip, and the gourd is filled with warm water. After it sits for some time, the insides are shaken out. The gourd continues to be filled with water and allowed to soak, until eventually the interior is completely removed. Occasionally, pebbles or rough-edged gravel is added to the water and shaken to help loosen stubborn pieces of pulp. Once the interior is cleaned out, the gourd is soaked in water to get rid of any odor, and then is allowed to completely dry in the sun.

DON'T FORGET TO
CLEAN UP THE GARDEN

It is easy to get so involved in the harvest and care of the gourds that you forget to do the finishing touches on the garden itself. This is a very important part of the process, since debris left in the garden area is a certain invitation for problems in the next season. Take some time now to clean up, and you will be rewarded with a much more trouble-free experience in the spring.

Remove the vines from the field or trellis completely. While many authorities recommend burning dried vines and bad gourds, many localities have laws against this. Therefore, bundle the vines and haul them to a dumpsite. Many bugs, insects and diseases overwinter in the debris and soil where gourds were grown. As a precaution for next year's crop, remove as much of the vegetation from the areas where the gourds of this season were grown, and if possible, till or turn over the soil. That way you prevent serious infestations for next year. You may want to begin rebuilding the nutrients that were removed by this years's crop by adding a layer of manure or compost, and then planting a cover crop, such as clover, cowpeas or vetch to replace valuable nitrogen.

SECTION III

Potential Problems and What To Do About Them

Gourd plants are susceptible to many of the same pests and diseases which affect the other members of the Cucurbitaceae family. While gardening guides and packaging on sprays or other controls do not mention gourd plants specifically, it is safe to assume that directions which are given for other squash and melons apply to the gourd plant as well. This chapter will describe in some detail the main problems which gourd growers are likely to encounter. There are many other potential problems as well, some that are specific to particular regions of the country, or that may be unique to your own setting. Make friends with your local nursery or garden center personnel. Many of these individuals are highly informed about the conditions and problems which are common in your particular locale. Different states have regulations which determine the chemicals which are allowed in their areas. A local nursery consultant will be able to help identify substitutes or solutions that have worked for others in your community. Pests and diseases do not honor property lines! Chances are that if your garden is bothered by something, neighbors have the same, or similar problem. Another excellent local source for information is the local Agricultural Extension Service. Frequently a telephone call will put you in touch with an expert who can answer your questions.

> *NOTE: **Do not experiment with chemicals in the garden**. Always read and follow the directions on the labels exactly. Many chemicals must not be used in combination with others—they will produce a reaction that can kill a plant. Many chemicals will also have deadly effects on the good visitors in the garden—honeybees, moths, bats, and 'good bugs.'*

Whenever you use chemicals in the garden, you should handle them with extreme caution. Even though they are not supposed to be harmful to humans, they are still poisons and should be treated with respect.

- Follow all directions on the container, especially safety precautions.

- Observe restrictions on number of applications and time of day recommended for use. (i.e. some are not to be applied in direct sunlight, or the leaves will burn).

Gourd plants, like most cucurbits, are very sensitive plants, so the mixtures should be the minimum strength recommended.

- Mix outdoors and wear a mask if possible.

- Use disposable surgical examining gloves when handling chemicals. Boxes of 100 are readily available at most pharmacy or surgical supply stores. Wash hands thoroughly when you are done.

- Chemicals used in various insecticides and fungicides often should not be combined or used on the same plant. Watch for this precaution on the labels of the products you are using.

Pests

Cucumber Beetle

The cucumber beetle is one of the most common pests in gourd gardens. Many gardeners are simply resigned to efforts to keep the population under control, since complete eradication is impossible. Charles Heiser, author of *The Gourd Book,* suggests that the cucumber beetle may even be one of the pollinators of the gourd plant, thus creating a real 'good news-bad news' dilemma. There are two types of cucumber beetle—one striped and one spotted.

Spotted cucumber beetle

The spotted cucumber beetle is also known as the "southern corn rootworm beetle." It is greenish yellow and has twelve black spots on the wing covers. There are actually two related species of striped cucumber beetle, one more prevalent in the eastern United States, and the other found primarily on the Pacific Coast. However, they are similar in appearance, life cycles and habits, so they are usually considered as a single pest. The striped cucumber beetle is yellow with black head and black stripes on each wing cover. Both the spotted and striped cucumber beetles are approximately 1/4 inch long.

The beetles spend the winter in the ground or in piles of rubbish such as old grass or plant debris. When the temperature warms, they emerge and immediately attack young plants from the Cucurbit family, feeding on the young leaves and stems near and below the soil surface. The plants usually curl and die.

Striped cucumber beetle
(American Phytopathological Society)

Mating begins immediately after the beetles emerge from the ground, and the females lay their eggs in cracks or crevices of soil near the stems of the plants. When the slender white larvae hatch they immediately crawl onto the plant and bore into the stems or roots. (The larvae can destroy the entire root system of a young plant; however an established plant may not be killed, but will suffer from dieback.)

The larvae then pupate, and the beetles emerge to feed on the plant stems, leaves and blossoms. The beetles continue feeding into July, feeding on many different plants, thus spreading diseases such as bacterial wilt and mosaic virus. By August the beetles tend to thin out, although they will not disappear altogether.

Cucumber Beetle Control:

- Till the soil in the spring to expose any hibernating adults. Some growers sprinkle the ground with Diazanon which is tilled into the soil. This rids the ground of any pests which may have overwintered.

- If seedlings are started indoors, they should be protected as soon as they are transplanted outdoors. The young plants can be dusted with an insecticide or they can be covered with a collar and netting, such as described on page 50.

- At the first appearance of the beetles, use one of the following chemicals: carbaryl (Sevin), malathion, or methoxychlor. Many people use Rotenone, applying 1-2 times a week once the plants emerge. Commercial growers often use a systemic insecticide Furadan, which provides some control early in the season.

- When the beetle population surges in June, it may be necessary to use several applications of insecticide spray, weekly through mid-July, when the population naturally subsides.

- Be aware of the potential problems that insecticides may cause, such as eliminating the good bugs and pollinators from the garden. Use chemicals with caution, and explore the other alternatives suggested on page 88.

Squash Bug

This pest excretes a foul-smelling juice and is commonly known as a "stink bug." The adults are over 1/2 inch long gray/brown and flattish in shape.

Like the cucumber beetle, they spend the winter in piles of rubbish or debris in the garden area. They emerge once the weather warms, usually in mid-June.

*Squash bug with eggs on
the underside of a gourd leaf
(American Phytopathological Society)*

As soon as they emerge, the bugs mate. The females lay clusters of shiny reddish-brown eggs on the underside of leaves, or on the stem of the plant. When the eggs hatch, they are gray spider-like nymphs. (They do not go through a larvae, caterpillar-like stage in their life cycle. They grow by shedding their skins, developing into adults in several stages. The nymphs vary from bright green with red head and legs, to dark greenish gray with black head and legs.) The nymphs start sucking the juice from the leaf where the eggs were attached, and soon infest the entire the plant. Besides causing leaves to wither they may puncture the skin of young fruit. This may not be noticeable as the gourd is growing, but once it is harvested it may rot instead of harden.

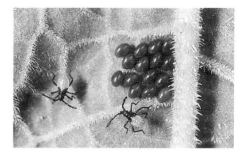

*Nymphs and eggs of the squash bug
(American Phytopathological Society)*

Squash Bug Control:

- The most direct control is to pick off any bugs or egg masses you see on the underside of the leaves. Inspect young plants regularly for this telltale sign.

- Try laying boards on the ground near the vines at night, since the adults usually gather in a protected place. In the morning, destroy any bugs you may find.

- Pesticides containing carbaryl (Sevin) have been found to be most effective. It is important to spray under the leaves, since this is where the eggs and nymphs are found. Insecticides are most effective against the small nymphs, so it is important to inspect plants carefully in early summer to control this pest.

Adult squash vine borer.
(American Phytopathological Society)

Larva inside the gourd vine
(American Phytopathological Society)

Squash Vine Borer

This relatively common gourd pest is the grub of a clear-winged moth (which looks much like a wasp.) It has prominent red-orange hairs on the abdomen and legs, making it fairly easy to identify. It tends to be more of a problem east of the Rockies.

The borer spends the winter in the soil as a full-grown larva in a tough silken cocoon as opposed to the cucumber beetle and squash bug which overwinter as adults. In the spring the larvae pupates and the moths appear by mid to late June. After mating eggs are laid on stems, leaf stalks and leaves within a foot or two of the root. As soon as the eggs hatch, tiny white grubs bore into the main stem of the plant, gradually working toward the root, although they seldom get below the soil surface. As they progress down the stem eating the vital parts of the plant, the leaves wilt and a section of the stem dies. If several grub are feeding on the same plant, it has little chance of survival.

Symptoms to watch for: Sudden wilting of the stem and leaves. Look for a small hole at the base of the stem with green 'sawdust-like' material in piles near the holes.

Squash Vine Borer Control:

• Examine the undersides of the leaves, and scrape off any egg clusters found on stems or the underside of leaves.

• If there is a sudden die-back of leaves, look for a small borer hole at the base of the stem. Split the stem lengthwise and look for the grub. Use an eyedropper or other small syringe to squirt an insecticide into the stem if you don't see the borer.

• Induce root growth beyond the point of injury by covering the vine with dirt. Most varieties of cucurbita will form new roots at the leaf joints, so that the plant will become reestablished. Once the plant seems revived, cut off and destroy the damaged stems. Unfortunately this cannot be done when the gourd vine is growing on a trellis.

- Try to prevent moths from laying their eggs by spreading shiny foil under the young plants. The sunlight reflecting on the undersurfaces of the leaves distracts and confuses the adults! Wrap a foil collar around the stem of the plant next to the ground for further protection.

- When the plants are young, spray with malathion or methoxychlor, paying special attention to the underside of the leaves.

Cutworms

There is a large variety of hairless larvae of night-flying moths that make up this diverse group. The larvae feed at night and on overcast days, and during the daylight hide curled in the ground. They can cut a new plant to the ground overnight.

Cutworm Controls:

- Many people till Diazanon into the soil before planting to kill any grub or larvae that may have overwintered in the ground.

- Seedlings should be protected by putting a physical barrier around them as soon as they sprout, or are transplanted into the ground. Simple barriers can be constructed of milk carton or cottage cheese container, with the bottoms removed. Coat the top edge with petroleum jelly or sticky ant barrier to prevent invasion from the top.

- Surround the base of the plant with a dusting of insecticide, or a sticky ant barrier.

- Place cardboard or wood slabs on the ground near the plants. During the day, lift them and destroy any grubs hiding underneath.

Aphids can quickly overwhelm gourd plants.

Aphids

There are many varieties of aphids that are attracted to different plants throughout the garden. The melon aphid is especially troublesome for the Cucurbit family. It is a small louse-like insect that sucks the juices from the leaves of the plants. Early in the season, small colonies form on the underside of the leaves, and they soon spread over the entire plant. They often cause the leaves to curl and eventually die off.

While aphids themselves cause damage to the plant, they pose a greater threat as the primary carrier of many viruses, such as mosaic virus.

Aphid Controls:

It is important to control the aphid population before it gets large. Often, frequent hosing, either with plain water or with a mild solution of insecticidal soap is sufficient to rid the leaves of the colonies. If this is not sufficient, you may need to apply an insecticide such as rotenone, lindane or malathion.

Diseases

The main types of diseases that affect plants are caused by three different organisms: virus, fungus and bacteria.

- *A virus* is an ultramicroscopic organism that invades the plant tissue, reproduces and threatens the life of the host. There is no cure for virus infections that is available for the home gardener, but you can take steps to reduce the chances of virus spreading to your plants. First, get rid of any plants that are stunted or mottled, or otherwise appear unhealthy. Second, control insects, such as aphids and cucumber beetle that carry viruses.

- *A fungus* is a multicellular organism which is a parasite that obtains food from green plants, thus causing additional diseases. They produce many spores, which can be carried by wind or water from leaf to leaf or plant to plant. Each spore, in turn, can germinate and grow, producing more infection. Many fungicides are available for the home gardener to control the more common fungi.

- *Bacterial diseases* are single celled micro-organisms which also get their nutrients from the host plant. They are most often spread among plants by insects.

What follows is a brief description of the main problems which gourd growers may encounter. It is by no means exhaustive, but your local nurseryman, extension agent or library should provide additional help if necessary.

Powdery mildew
(American Phytopathological Society)

Powdery Mildew

This is a fungus on the leaves that usually appears in the late summer. The grayish white powder first appears as small spots on the surfaces of the leaves, which gradually increase to cover not only the leaves but the stems of the plant as well. The leaves eventually wilt and die.

Powdery mildew is commonly thought to require moisture to grow. However, it is caused by a spore that is carried in the wind. Moderate climates, cool evenings, shade and crowded plantings make ideal conditions. Although humidity and moisture can stimulate growth and spreading, the spores need dry leaves to become established.

The disease will overwinter in litter or debris that is left on the ground, or on perennial plants that continue to live in the vicinity after the garden area has been cleared. It will first appear on plants as yellow spots on the older leaves. As the spots grow, they develop characteristic powdery film that eventually covers the entire leaf.

Powdery Mildew Control:

- Prune out infected leaves, and remove from the garden area.

- Many fungicides are on the market to control mildew. Some products are designed as a preventative, while others aim to get rid of the mildew once it appears. Most products for powdery mildew contain sulfur: karathane, sulphur benomyl, or a bordeaux mixture (copper sulfate and hydrated lime).

- A home remedy can be made using a mixture of baking soda and mineral oil to control mildew. Mix 2 teaspoons of baking soda and two teaspoons of lightweight horticultural spray oil (Ultrafine Sunspray) with one gallon of water. Be sure to spray all leaf surfaces for effective control.

Anthracnose

This is another fungus that is caused by an organism that frequently thrives in the soil and can affect many plants and even trees in the yard. It can be transmitted to a healthy plant by raindrops that splash contaminated soil on the leaves. Wind can also blow spores from one infected area to plants. Some varieties of gourds tend to be more susceptible to anthracnose than others: penguin, birdhouse, cannonball and basketball seem to be particularly susceptible. Those less affected are maranka, dipper, Indonesian bottle and ornamentals.

Anthracnose
(American Phytopathological Society)

Anthracnose appears as large, irregular brown blotches on the leaves. Dry centers of the blotches may fall out, leaving a "shot hole" appearance. When it gets on the fruits it causes lesions and scabby patches. Young fruit may turn black, shrivel and die. The spores may invade the inside of the fruit, where they will stick to the seeds. (It is therefore very important to get rid of infected gourds so that the fungus won't be perpetuated by future generations.)

Humid or wet weather favors the disease. Plants can become affected at any stage of their growth, but the most frequent appearance is mid to late season after the vines and fruits are formed.

Anthracnose Control:

- It is very difficult to control anthracnose once it is established. The most important preventive measure is to rotate crops at least every three years.

- Spores can survive over the winter, so it is important to remove all diseased material and destroy it. The smallest amount of infected material can restart an infestation.

- Control weeds in the area of the gourd garden, and try to avoid working with the plants when they are wet.

- Many fungicides are available to the home gardener today, although anthracnose is particularly resistant to control. The best program is preventive.

Begin the application of a fungicide spray program right after the first true leaves form. Spray weekly in warm rainy weather until just before harvest. Suggested fungicides include rotenone and other sprays which include copper solutions.

A garden suffering from Bacterial Wilt.
(American Phytopathological Society)

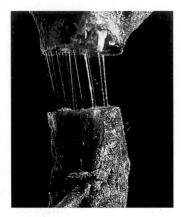

Bacterial Wilt can be identified by the sticky substannce in a cut stem. (American Phytopathological Society)

Bacterial Wilt

This disease is carried by the cucumber beetle. The plant wilts because the bacteria plugs the water ducts of the stems. Bacterial wilt can be distinguished from other wilting by a white, sticky, stringy substance (bacterial slime) that can be seen when a stem is sliced and pulled apart.

In the early stages of the disease, the leaves may wilt in the daytime and recover at night. (This may appear to be a problem of the plants needing more water. If you have irrigated sufficiently, or if the ground is moist, suspect bacterial wilt.) Young plants will die rapidly, but older plants may be affected on one branch, and the rest of the plant not be affected at first. Wilt also causes fruit to wither and drop off.

Bacterial Wilt Control:

Because the disease is spread by the cucumber beetle, you can greatly reduce the threat of this problem by controlling the cucumber beetle population. (see above description)

Mosaic Viruses

There are several related viruses that are spread by aphids and other insects like the cucumber beetle, attacking the vascular system of the plant. They first appear as yellowish green mottled leaves, stunted vines, which produce warty deformed fruits. New leaves may wither and die, and the vine eventually dies back.

Mosaic Virus Control:

Because there is no direct cure of mosaic viruses available for the home gardener, the only reliable defense is to get rid of the aphids and cucumber beetles that spread the disease. Spray with malathion or other insecticides recommended for aphid control throughout the growing season.

Good Garden Maintenance

Prevention is the Best Cure

Given the variety of problems that may beset a home gourd grower, the best steps to take for a relatively relaxing and enjoyable gardening experience are prevention. Here are some recommendations on simple ways to prevent some of the more serious problems before they start, or at least slow them down if they do make an appearance in the garden.

- Clean up the entire garden area as early as possible, preferably in the autumn, after the last gourd is harvested. Remove all vines by burning or by bagging and carting to the refuse area.

- Till the soil in the autumn to disturb or expose any bugs that may have settled down for a long winter's nap. Add compost and manure to replenish the soil, and if possible, plant a cover crop.

- In the spring, till the soil again to expose any bugs or larvae that may be in the soil. Many gardeners choose to till in Diazanon to kill grubs in the soil. Be aware that insecticides/pesticides are not choosy—they will kill good insects in the soil as well. So use chemicals only if you know there is a problem with insect pests in your area.

- Plan a watering system that does not wet the leaves or sprinkle so that contaminants from the soil can bounce up and infect the leaves. (This applies especially to soil borne fungal diseases.)

- When seedlings begin to emerge from the ground, provide a barrier such as a milk carton, as protection against bugs and larvae.

- Use hydrated lime (less than 25% magnesium) and woodash around rows or hillocks of new plants, to prevent rootworm and many insects. This is one of the safest chemicals to use. Put it on the soil before planting for long-lasting control of many insects and diseases. There are two cautions to be aware of before using either lime or woodash:

First, lime is traditionally used to adjust the pH levels of overly acid soils. It adds calcium to the soil, and may be used to replenish nutrients to the soil as well as act as a general purpose insect preventative. Before using lime in any quantify, check the pH level of the soil. Keep in mind that gourds prefer a pH level between 6.0 and 6.5—that is, slightly acidic.

Second, woodash is very high in potassium, so be careful not to add too much ash to the garden area and completely upset the soil composition.

- Begin a program of spraying or dusting plants as soon as plants emerge from the ground. This can be continued on a weekly basis through the middle of July. Once the gourds are set, less frequent applications should be sufficient. Readymade dusts in garden supply stores frequently combine insecticides and fungicides.

Common Remedies

Natural alternatives

Before resorting to the use of chemical warfare against your garden invaders, you may want to try other methods of control first:

- Wash off eggs and aphids with a strong spray of the hose. If plain water does not solve the problem, try a very light application of insecticidal soap. This will not be harmful to the beneficial insects in your garden.

- Pick off all eggs, grubs, and larvae. This is difficult once the plants begin to vine and grow rampant. But if you are conscientious at the beginning of the season, you may find that a small problem never becomes a big one.

- Use barriers around seedlings, and spread a thin layer of ashes and lime around the base of new plants.

Many newer books and literature talk about the advantages of companion planting. One year I was very concerned about the amount of chemicals I was applying to my plants and the ground, and vowed to try a different method the following year. Between the gourd plants I added seeds and bedding seedlings of all the plants that the literature recommended as possible deterrents. The results were spectacular—it was the most worry-free year I have had in the gourd garden. The best part of this experiment is that you have nothing to lose—if it works, you and your garden are winners! If it doesn't work, you have lost nothing, and can still use other controls listed above. Companion plants are effective primarily against insect pests; they offer no protection against diseases, other than reduce the population of the insects that spread them. However, with gourds insects *are* the agents of contamination in most instances, so the protection is an important consideration.

- **Radishes:** These are effective primarily against striped cucumber beetles. Plant seeds throughout the gourd garden. Harvest some radishes for salads, but leave as many plants as you can to go to seed. They are effective throughout the growing season.

- **Catnip:** Plant throughout the garden, either from seed or from bedding plants available at most nurseries. Your cat will love you (even more!), and you will eliminate (or at least reduce) aphids, squash bugs and cucumber beetles.

- **Broccoli** is thought to be an especially effective repellent against cucumber beetles, and therefore a protection against bacterial wilt.

- **Tansy:** Add this herb to repel aphids and squash bugs.

- **Dill:** May repel aphids, and is always a wonderful addition to garden salads.

- **Marigolds:** In addition to providing a lovely color contrast at the base of the gourd arbors, marigold roots create a chemical that kills many nematodes (insects in the soil, such as cutworms). They produce the chemical slowly, so if left in the garden for the entire season they will provide control for up to three years.

- **Buffalo Gourd** (coyote gourd, bitter gourd): These are wild relatives of the domesticated cucurbits which are normally grown by the home gardener, native to the arid Southwest. Elders of the Cochiti tribe have passed on advice for their use as a natural pesticide. While all gourds contain cucurbitacins (a foul smelling chemical that is extremely bitter to the taste), the buffalo gourd appears to have a particularly strong concentration of it, which is reputed to repel many insects and animals. To use in the garden, crush the gourd in water, and then sprinkle on the gourd plants to repel squash bugs. The elders also advise their people to cut the gourd and some leaves to hang in the corners of their homes to keep out insects!

Marigolds can provide both color and protection to your gourd garden.

Many other plants, such as flowers, are valuable additions to the gourd garden for attracting beneficial insects. Often pests in the garden can be controlled by natural predators, thereby reducing the need for chemicals. In planning your gourd garden, consider adding a variety of other plants to attract a wider variety of insects to the area.

Chemicals

If you do choose to use chemicals as a means of insect and disease control, there are many alternatives from which to choose. Some of the more common ones that are available for the home gardener are described below.

Insecticides—what they are, how to use them.

- **Rotenone**: This is a botanically derived insecticide that is obtained from 65 plant species. It can be toxic to mammals and is extremely toxic to fish, but does not affect humans. This insecticide is most frequently recommended to control chewing insects.

- **Pyrethrum**: This natural insectide was originally derived from pyrethrum daisies, but is also found in many species of chrysanthemums. Some forms of this insecticide which are specifically derived from the daisy will break down after exposure to sunlight. It also comes in synthetic versions that are more toxic and last longer. It is effective against many insects, including beneficial ones. This spray is also highly toxic to fish and frogs, so don't spray near ponds.

- **Insecticidal Soaps:** These are mixtures of special fatty acids that are not harmful to humans but control most small insects, including aphids. You can buy a commercial product, or make your own by mixing 1-3 teaspoons liquid dish soap in one gallon of water. Spray cautiously, however, for repeated frequent use may actually harm cucurbits.

- **Diazanon**: This insecticide is the only chemical control for soil pests in vegetable gardens. If sprayed, it will kill insects, and is highly toxic to bees and birds. Therefore, use with caution!

- **Carbaryl (Sevin):** This insecticide is available in dust and spray forms. It is most commonly recommended to control a wide range of pests in the vegetable garden. It is most effective against chewing insects, but not sucking ones. Sevin is highly toxic to honeybees and earthworms, so it should be used with caution.

- **Neem**: This chemical is extracted from the seeds of the Neem tree found in India. It is effective against cucumber beetles and many other insects, as well as preventing some fungal diseases. It seems to have no toxic effect on humans, although it still should be used with discretion.

- **Oils**: Special, highly refined oils smother insects and their eggs. They are especially effective to control aphids, mites, and the eggs of other insects. Some oil sprays are also effective against fungal diseases.

- **Parasitic nematodes**: There are many different kinds of nematodes. The kinds that are used to control garden pests are extremely tiny, and carry a bacteria, which, when eaten by insects, multiply and kill the insect. Nematodes are used primarily to control soil pests or those that live in protected spaces such as under leaves and in mulch. This would include the cucumber beetle, squash vine borers, squash bug, and other grub or rootworms. Nematodes are usually sold as a powder, which is mixed with water and either sprinkled or sprayed on the soil.

Some other products which are frequently recommended by growers or mentioned in gardening literature are not available in all states or localities. They are Dursban (chlorpyrifos, Kethane (dicofol), endosulfan, and nicotine.

Keep in mind that there are reasons these products have been limited from many areas, and you may not want them in your garden either. A frank talk with your local garden expert will advise you on what is available in your locale.

Disease controls—what they are, how to use them.

- **Oils**: Make your own mixture by combining ultrafine gardening oil and baking soda for control of powdery mildew, anthracnose, and other fungal diseases. This is also an effective insect control, and the oil smothers eggs before they hatch.

- **Lime-sulfur** (calcium polysulfide): preventive against powdery mildew, some spider mites and small insects.

- **Sulfur**: One of the oldest fungicides, commonly used to prevent powdery mildew, scab, and other fungi.

- **Copper compounds**: This is a group of general purpose fungicides used to prevent mildew and anthracnose, among other garden problems.

LOCAL HELP

Help may be as near as your telephone. Most local governments have a Department of Agriculture which is available to provide advice and support for the home gardener. Check the phone book under government listings, starting at the county level. Most counties have a Commission of Agriculture, staffed by biologists or other professionals who may provide suggestions either by phone, in their office, or at your home. In my area, the County Agriculture Commission has a Master Gardener program. These highly informed individuals will give advice concerning soil, planting zones, etc. If you are troubled by specific pests or diseases, they may suggest that you take samples to the county office for identification and suggestions for treatment. Agriculture Extension agents are also employed by all state governments and are very knowledgeable about local agricultural issues. These professionals are eager to help you enjoy a successful gardening experience—take advantage of this excellent opportunity in your own back yard!

SECTION IV

Uses for Your Gourds

People grow gourds for many reasons. Some like to watch the lush vines spread dramatically over a corner of the garden, trained along a fence or trellis, or covering a field. The white blossoms of the hardshell gourds come out as brilliant stars in the evening against the sea of dark green leaves. Other gardeners are intrigued by the endless possibilities of crop experimentation, either hand pollinating or otherwise manipulating the new crop.

The ultimate reward for the gardener's efforts is an abundant crop of fanciful and fascinating gourds. And many growers, I am sure, stand in their fields in the fall, hands on hips, and wonder "What can I ever do with all of this?"

As mentioned in the first section, humans have been using gourds for many different purposes for thousands of years. In many parts of the world, gourds are still an important crop for a village or a society. Gourds have been, and still are, grown for food. Different types of gourds are favored by different cultures, and it is quite interesting to explore new tasty surprises from around the world without ever leaving your own back yard.

Gourds have been used as containers for every possible use one might imagine, from animal houses to every manner of domestic, ceremonial and ritual purposes. Other books are available which describe specific craft techniques in detail, and there is not room in this book to cover the possibilities available today. This section will serve to introduce a few of the basic ideas of how you might put gourds to use in your own life.

Birdhouses at the Museum of Appalachia, Norris, Tennessee, attract large colonies of Purple Martins

Luffa and Tomato
Ratatouille

1 medium onion, sliced
1 garlic clove, crushed
2 Tbs. cooking oil
1 lb (450g) sliced luffa
2 large tomatoes, skinned and
     sliced
salt and freshly ground pepper
6 large leaves fresh basil, shredded
béchamel sauce (optional)
grated cheese (optional)

Soften the onion and garlic in a
saucepan in 2 tablespoons cooking
oil. When transparent add the
sliced luffa and simmer until soft.
Add tomatoes, seasoning and basil
and simmer until all ingredients are
soft. Add a little water if necessary
so that the mixture cooks evenly.

Optional: Spoon the ratatouille
into a small ovenproof dish, cover
with béchamel sauce, sprinkle with
grated cheese and brown in the
oven or under the grill.

Recipe from *Oriental Vegatables:
The Complete Guide for Garden
and Kitchen*, by Joy Larcom

GOURDS FOR FOOD

Originally gourds were highly valued by people for two important features: for the unique woody shell, and also as a food source. The very large Cucurbitaceae family includes many members which have been cultivated since ancient times for food, and still are consumed to this day. We are all familiar with the squashes, pumpkins and melons that fill the markets and cookbooks. A trip to a market in every foreign country will introduce many new and unusual varieties that most of us have never encountered in our local markets. These fruits and the ways they are prepared vary greatly among cultures, and can provide an exciting culinary adventure.

This section will focus primarily on the ways other cultures have prepared the gourds that are dealt with in this book—primarily the hardshell gourd and the luffa, as well as some of the gourds mentioned in the appendix.

The hardshell gourd (Lagenaria) provides both a thick shell and a huge number of seeds, both of which have been used as food source. The seeds contain a relatively high amount of oil, and have been prepared in various forms: raw, shelled and roasted, and ground as a mush. Gourd seeds were most commonly eaten in the Middle East and the Mediterranean area, although they were also eaten by the Native American Indians, Hawaiians, and even Maoris.

I have tried eating gourd seeds, both raw and lightly toasted. It is very difficult to get the protective shell off. (It makes you appreciate why the gourd seed is such a slow starter in growing!) A small pliers works best to crack the shell. Once that is off, however, the seeds had a bland flavor.

I did not detect a noticeable difference between seeds of the varieties I tried (canteen, Indonesian, bottle, and bird-house. Other varieties, such as Maranka, may have different flavors) A word of caution, however: I have read that some gourd seeds contain a very strong, bitter tasting heart stimulant. (colycinphine) One recommendation is that you munch the gourd seeds slowly, and in moderation. If any have a bitter taste, throw them out.

More frequently, the rind of the gourd is the valued part, particularly that of the small, immature gourd. Pick the hardshell gourd while it is still small (relative to the expected mature size). This will vary among varieties: (a canteen gourd could be picked at 2-3" diameter, and a bottle gourd when it reaches 4-8" in length. Glenn Burkhalter described how a visitor to his gourd farm tested the hardshell gourds with her fingernail. If the fingernail could penetrate easily, it was suitable for eating. It can then be sliced, steamed and seasoned much like a zucchini or other summer squash. The flavor is bland, so you may want to experiment with seasonings or adding other ingredients. Because young gourds are frequently eaten in Asia and India, many recipes for them can be found in specialty cookbooks or markets. The Italian cucuzzi, a hardshell gourd that is commonly referred to as a snake gourd, is specifically grown in many areas as a vegetable. When it is small, it looks like a cross between a long green bean and a small zucchini, and is quite tasty. It can be prepared using many recipes you might have for summer squash or zucchini. Again, it is important to harvest the vegetable while it is small.

Indonesian Gourd Stuffing (Petola Daging)

2 lb (1 kg)) luffa or other gourd, thinly peeled
4 oz (100 g) shelled prawns
8 oz (225 g) minced lean beef or chicken breast
4 eggs
salt and freshly ground pepper
vegetable oil
1 medium onion or 2 shallots, thinly sliced
2 green chilies, deseeded, or 1/2 teaspoon chili powder

Halve the luffa or other gourd lengthwise and remove seeds. Leave in cold salted water while preparing the stuffing

Chop the prawns and mix with the minced meat. Separate one egg, keeping the white to one side. Mix the yolk with the other whole eggs, season, and cook to make several thin omelets. When cool, roll up and cut crosswise into thin slices.

Heat a little oil in a wok or pan, cook the onion or shallots and the chili until just soft. Add the prawn and meat mixture and stir-fry for 3 minutes. Season, cool slightly and stir in the reserved egg white. Mix in the omelet pieces.

Drain and dry the luffa or gourd halves. Stuff the bottom half, and use the top half as a lid. Steam for 30 minutes or bake for 40 minutes in a moderate oven (350° F). Cut into thick slices. Eat hot or cold.

Recipe from *Indonesian Food and Cookery,* by Sri Owen.

Loki Kofta

2 cups grated edible gourd,
   (called calabash in Oriental
   grocery stores or loki in
   Hindi)
1 cup gram flour (chickpea flour)
1/2 cup cornmeal
4 cloves garlic, minced
1 tsp. fresh ginger, minced
1 tsp. green chili or to taste
1/2 tsp. salt
1/4 cup oil

Mix together everything but the grated gourd, then gradually combine the dry ingredients with the gourd. Form into small balls; these are the kofta. Heat oil to medium high, add kofta and fry until golden brown.

Recipe from *The Gourd,*
Fall 1988.

The Japanese have a special use for the rind of the mature bottle gourd. It is cut into strips and dried; packages of them are then sold in stores as *Kampyo*. The packages contain recipe suggestions, although most frequently it is one of several ingredients in a dish. It can be used in a soup, stir fry, or sushi, among other preparations.

The Chinese also have a drink which is made from a gourd, which tastes similar to sweet tea. It is frequently available in cans in Chinese food stores.

The most commonly used gourd in foods today is the luffa. Visiting farmers' markets you will frequently see piles of immature luffa, usually the *Luffa acutangula* (known in Chinese as *singua*). The smaller luffas are more flavorful, so select ones that are no more than 10-12 inches long. Once the outer skin is peeled off, the inside is soft and slightly spongy. The simplest way to prepare this is as a stir-fry, cooking it only briefly along with other vegetables and perhaps seafood. In India, luffas are frequently used in curries.

Farmers' markets or specialty food stores may not be familiar with our term "gourd." Other languages refer to "gourd" as: *houtain yugao* (Japanese), *kampyo* (dried gourd rind) (Japanese), *po gua* (Cantonese Chinese), *kwa kwa* or *sing kwa* (Chinese), *upo* (Filipino), *bao* (Vietnamese), *dudhi, lauki* or *loki* (Indian).

GOURDS AS MEDICINE

Anyone who has worked with gourdcraft will recognize that there is something unusual about this plant. The dust of the pulp may result in an allergic reaction in some people, causing a swelling in the mucous membranes. This may be experienced as swelling or itching eyes, strange feeling in the mouth, and constrictions in the breathing passages—almost an asthmatic condition. (For these reasons we always recommend that crafters wear at least a breathing mask when cleaning or carving gourds.) You may also notice a strange bitter taste on your lips and hands. This is due to *cucurbitacins*, which are present to some degree in all cucurbits but are particularly concentrated in some species and varieties. (These cucurbitacins act as an appetite stimulant to cucumber beetle and squash bug, but can have quite the opposite effect on humans.) Another active ingredient that has been identified as particularly concentrated in some gourds is *colycinphin* a very strong heart stimulant. Very little serious study has been done on these ingredients and their actual effects on humans.

The Healing Gourd

In China today the bottle gourd is recognized as a symbol of herbal medicine and healing. The origin of this symbol comes from an ancient legend which describes Li-Tiequai, one of the Eight Immortal Gods, who is the patron deity of medicine. He is most frequently depicted as a crippled beggar with a crutch in his hand and a bottle gourd tied to his sash. According to legend, however, he began as a handsome young noble who studied philosophy, striving for immortality. One day he instructed a young disciple to watch over his body as his spirit went to visit Lao-Tze at the Sacred Mountain. When his spirit returned to his body many days later, he found that it had been burned to ashes. Anxious to find another physical form, Li's spirit entered the body of a beggar who had recently died. When he realized his new misshapen form, he pleaded with the gods for help. Lao-Tze responded by giving him a crutch and a bottle gourd of magic herbs to cure mortals. After that time, by day he wandered the streets as a healing mendicant, and at night his spirit entered the bottle gourd which contained a magical world of comfort and peace.

97

We do know that the gourd plant has been used extensively in folk medicines in cultures around the world to treat an extremely wide variety of ailments. The first recorded description of the use of gourd plants for medical purposes appeared in the works of the Roman naturalist Pliny, writing in the First Century AD.

Other similar remedies appear in cultures around the world. Many of the gourds specified were not the hardshell gourd but relatives such as the balsam pear (momordica charantia), the Chinese snake gourd (trichosanthes), or the bitter gourd (colocynth). The reader is referred to *The Gourd Book*, by Charles Heiser, for a more complete description of these gourds and their medical uses.

A species that is more closely related to the gourds featured in this book is the buffalo gourd, or coyote gourd (Cucurbita foetidissima), which is a wild gourd native to North America, found primarily throughout the southern Plains region and the Southwest. Every part of this remarkable plant was used by the Native Americans for healing.

The fruit was cut open and rubbed on wounds on animals and humans. The leaves were steeped, mashed, and turned into a poultice. This was supposed to ease swelling in muscles, or even swelling caused by mumps. The poultice could also be applied to the forehead to provide relief from a headache. The roots of this plant

The Roman Naturalist Plini Described Some of the Medical uses of Gourds.

- Use as a purgative, for all diseases of the intestine, kidneys and loins.
- Apply a combination of pulp with wormwood and salt to cure a toothache.
- Juice made from the pulp warmed with vinegar strengthens loose teeth.
- Juice is also good for earaches.
- Rubbed in with oil removes pain of spine, loins and hips.
- The inner flesh of the green gourd is good for corns on the feet.

were ground into a powder and mixed with water to create a cure for many different ailments: if swallowed it acted as a laxative; as a poultice it cured an earache; poured into a tooth it would stop a toothache.

Native Americans on the East Coast of America made use of the hardshell gourd for many of the same purposes that the Indians in the Southwest used the buffalo gourd. The leaves were steeped and crushed to be used as a general purpose healing poultice. The seeds were boiled to make a cure for the chills. Early settlers learned many of these secrets from the Indians and incorporated them into their home remedies as well.

The luffa plant was widely used for medical purposes in many Asian countries, where the plant probably originated. The seeds were used as a purgative. The leaves were made into a poultice to cure certain skin diseases. In China, the young fruit was thought to be beneficial to the stomach and intestines, and a tonic to the genital organs!

Given the historical uses of the gourd plant for healing purposes, it is quite surprising that little if any serious investigation has been undertaken to examine the plant more closely. Certainly there are some ingredients that have been recognized and put to work by our ancestors. Often when folk remedies are scrutinized, active ingredients are found which provide an explanation for their widespread use. Hopefully someone will be piqued by this great gap of knowledge and will be challenged to examine the Cucurbit family more closely. Most of us already recognize that the gourd possesses magical qualities. We welcome greater understanding in this area!

*This gourd was home to
a family of wrens*

BIRDHOUSES

Gourds provide a wonderful material for homes for birds, regardless of where you live. While a great deal is written to promote gourd houses for the purple martin population, gourds may be adopted by any cavity-nesting bird.

It is important to be familiar with the varieties of birds in your area, their size and nesting patterns, before you plan your birdhouse. The size and shape of gourd, the size and location of hole, and the eventual placement of the birdhouse vary among bird species. The following chart will provide general guidelines for size of the gourd and the size of hole for each species.

The shape of the gourd is not as critical as the internal diameter of the gourd bulb. If the bird has enough room to turn around once inside the gourd, the other dimensions (i.e. height) are not significant. While some gourds are classified and described as "birdhouse gourds," you actually have a wide choice in shape of the gourd.

Once a gourd has been selected with an appropriate size cavity, the next consideration is the placement of hole. For most bird species it should be at least 4 inches from the base of the gourd. Cavity dwelling birds build up a nest in the base of the gourd, and prefer to hop down to the nesting level. This also offers some protection for fledglings, so they will not topple out of the nest prematurely.

The size of the hole is critical: holes that are too large will permit larger birds to invade a nesting site. Holes should be cut precisely, and smoothed so no jagged edges protrude. You can get drills to make holes of the exact diameter desired, from 3/4 to 2-1/2 inches. These are very helpful, but not essential; you can saw the holes by hand, taking care to remove uneven edges and keep size accurate.

The insides of the birdhouse should be cleaned as completely as possible, although this can be a difficult task given the small opening. Using a wire, barbecue fork or prong, break up and remove as much of the pulp as possible. For stubborn innards, soak the gourd in warm water for several hours, add a handful of medium size gravel and shake briskly. The birds will clean out any remaining debris that is in their way.

At least four drain holes between 1/8 to 1/4 inch should be made in the bottom of the gourd. In addition, two holes should be drilled in the top for a thong or wire for hanging the birdhouse.

Bird	Minimum Size Gourd Needed Inches in Diameter	Exact Size Entrance Hole in Inches
House or Berwick' s Wren	4	1
Carolina Wren	5	1 - 1/8
Chickadee	5	1 - 1/8
Tufted Titmouse	5	1 - 1/4
Downey Woodpecker	5	1 - 1/4
White Breasted Nuthatch	5	1 - 1/4
Small Owls	5	1 - 1/2
Bluebird	5	1 - 1/2
Tree Swallow	5	1 - 1/2
Hairy Woodpecker	5	1 - 1/2
Crested Flycatcher	6	2
Flicker	7	2 - 1/2
Purple Martin	8 - 13	2 - 1/2

These are the only essential elements for a successful bird-house. Several other options may be considered:

Some bird varieties will use a perch, but many birds simply fly directly into the hole. Many people add perches with twigs, gourd stems and dowels. These can be added by lacing to the front of the gourd, or by drilling a hole and securing the perch with wood dough, putty or carpenter's glue. (Such fastenings may eventually loosen in time. This is due to the gourd shell becoming slightly bloated when wet, and uneven expansion between the gourd shell and the adhesive compound during hot and cold temperature variations. This joint should be examined periodically and reinforced by an additional injection of glue or putty.)

The birdhouse will last longer if it is treated with a wood preservative, such as Thompson's Waterseal. Such products are available at the hardware or paint store. If you are making many bird houses, a satisfactory and cheaper alternative is to use copper sulfate, available at a garden supply or hardware store. Dip the gourds into a solution of 1 lb copper sulfate/5 gallons of water, drain and allow to dry for several days before further work. This dilute solution does not appear to harm either birds or fledglings, and will definitely prolong the life of the gourd.

Because untreated gourds will absorb moisture and soften when left outdoors in the weather, it is a good idea to coat the outside of the gourd with several coats of enamel paint, varnish or shellac. If the birdhouse is to be hung in the open where it is exposed to direct sun, as is recommended for purple martin houses, paint the gourd white to reflect the sun and reduce the interior heat. Birdhouses that are hung in the tree canopy or in other sheltered locations do not need to be painted, but should be varnished.

The materials used to hang the birdhouse should be sufficiently strong to support the weight of the birds and nest, as well as the gourd. Because of the abrasion caused by wind and the movement of birds in and out of the nest, many leather tongs and jute cordage will break. Rawhide, polypropylene line or insulated wire work very well for this purpose. The hole through which the thong passes should be drilled to make sure the house is balanced. Drill the holes for the hanging so that the birdhouse will naturally sway forward and backward (holding the gourd so that the hole is in front.) This movement will naturally give when the bird flies into the hole. If you place the holes such that the birdhouse sways from side to side, it will be difficult for the bird to land on a breezy day!

Placement of the birdhouse is a critical factor in attracting nesting birds. For instance, wrens prefer a wooded bushy area such as tree canopies; bluebirds prefer a semi-open area, such as on a fencepost or a thin tree near an open meadow area; and purple martins select gourds that are entirely in the open. Some birds are very territorial and once established in an area, will chase away other birds, especially of their own species. Many birds will tolerate other species in their territory, but not immediately close by. Learn about the nesting habits of the birds you wish to attract before setting out the gourds.

The greatest challenge is attracting the first tenant. Once birds have nested in an area, they will often return to that site in subsequent years. After a gourd has been used for a season, it should be taken down and cleaned before the next season. There is some controversy surrounding this issue of cleaning out the gourd; however, parasites may be in the old nest and can live through a winter season. While the birds may not insist on a clean house in the spring, it is safer to clean out the old nest prior to storage for the winter. The exterior can then be repainted or revarnished to insure a long life.

Purple martins enjoy a very special relationship with mankind which goes back centuries. The earliest explorers in North America reported seeing gourds hung in many of the native villages and around their cultivated plots in the Southeast. The Indians had long recognized the value of purple martins to chase away other birds which eat the seeds or damage crops, and also to control insects around the villages. Settlers adopted this practice of erecting gourd houses for the purple martins. After centuries of such treatment these birds have become completely dependent on man to provide nesting sites. The

Purple Martin houses should be placed in an open area.

*Bird feeders can be hung
or mounted to provide
a protective perch.*

Purple Martin Conservation Association (Edinboro University of Penn., Edinboro PA 16444, (814) 734-4420) is devoted to the study of these birds and informing the public in how to keep the bird population healthy and strong. Their migratory patterns are well documented, and people from Florida to New England, east of the Rockies, are kept informed when to set out their nests, since the birds usually return to former nesting sites. Purple martins are social birds, and nest in large colonies. Therefore, set out many gourds—up to 24—to accommodate many nesting pairs. They are particular about location, however. Gourds should be hung 10 to 14 feet high, close together but with the openings facing slightly different angles. Because purple martins have been semi-domesticated, they prefer housing sites that are near houses or buildings, but in an open space. Therefore, gourds should be located within 40 to 100 feet of a house and in an open area that is at least 40 to 50 feet from any obstruction to provide ample room for foraging for insects and to allow for aerial flight directly into the nest. Additional information regarding attracting and maintaining a purple martin colony can be obtained from the above association.

Bird Feeders

In addition to birdhouses, gourds make very satisfactory bird feeders. The requirements for a bird feeder are not as rigorous as for a birdhouse. The gourd should have a relatively flat bottom so as to hold seed as well as perching area for one or more birds. This base should have several small drain holes in case of rain. The size and number of openings are entirely up to the gourd artist. Perches, decorations, and hanging location are also a matter of personal choice.

Bird feeders can be hung in trees or from wires, mounted on a flat surface or set on a permanent pole. If squirrels are a problem, the feeder can be suspended with fishing line from a wire or pole. If the gourd is placed in a location where winds might cause it to tip and spill the seed, fishing weights can be suspended from the drain holes in the bottom to reduce the sway. The bird feeder can be decorated or embellished as desired. It should, however, be coated with several coats of varnish or polyurethane to protect it from the weather.

SIMPLE IDEAS
FOR GOURD CRAFTING

Once you have successfully raised, harvested and dried a gourd crop, you may now survey your success with a bewildered, "Now what?" The scope of this book is not to present an in-depth look at gourd craft. For further information in this area, there are several books that cover this material. A current book on this topic is *The Complete Book of Gourd Craft*, which presents a comprehensive discussion of most of the techniques used for gourd embellishment today.

Gourd containers can be used for all types of snacks.

There are many simple but effective ideas to experiment with before you take the plunge into more elaborate crafts.

Cleaning a gourd

Unless special steps were taken to scrape the epidermis off the gourd while it was still drying, the dried gourd will be somewhat dirty and covered with a peeling, flaking or slightly molded outer layer. This is the waxy epidermis that protected the gourd as it was growing but no longer has a purpose. To get rid of this, first soak the gourd in warm water for up to one hour. Because gourds float, it is difficult to put a gourd in water and have the overall surface soaked. Many methods have been devised to simplify this process:

Use a copper scrub pad to clean the soaked gourd shell.

- Put the gourd (or gourds) in a bucket of water and then submerge them with a weighted cover.

- Put the gourds in a gunny sack that is weighted with bricks, and place this in a bucket of water. The bricks will keep the gourds submerged so they can soak evenly.

- Leave the gourds outside in the rain, or place them strategically under a sprinkler system.

- Put the gourds in a large plastic bag, add some water and tie it closed. Leave the bag in the sun, rotating it occasionally so that all the gourds are dampened. The warm water generated by a sunny day will soften the epidermis quickly.

- If there are stubborn spots of epidermis on the shell, try squirting it with a household cleanser and allow to stand for several minutes. The epidermis should wash away easily.

While some people put detergent or a household disinfectant in the water while the gourd are soaking, it is not necessary. Once the gourds have soaked for a period of time (10 minutes or more, depending on the condition of the mold), remove them one by one and scrape off the exterior with a kitchen scrubbing pad. Both plastic and metal pads can be used; however, the copper pad that is available in most kitchen supply sections of grocery and hardware stores is probably the best. It is abrasive enough to remove all but the most stubborn spots, and yet soft so that it does not leave scratches on the surface of the gourd. Rinse the gourds and allow to dry.

Note: Frequently articles in popular literature suggest that you scrape the epidermis off a dry gourd with a dull knife or steel wool. This will cause minute scratches in the gourd shell which may interfere with your designs. Always soak the gourd in water before you clean the shell.

Polished gourds

With many gourds it is not necessary to do anything at all for embellishment—the shape, texture and subtle shadings created by the molded epidermis combine to create a complete work of art without any outside interference. You may want to provide a finish to seal the porous woody shell and protect it from weather or accidental stain. A coat of wax may be all that is necessary. Any wax that is formulated for floors, furniture, wood or leather is satisfactory on a gourd shell. Some of the waxes that are available in houseware or hardware stores may be tinted or colored. These dyes or pigments will enhance the natural mottling that already exists on the gourd shell. Several coats of wax may be necessary to create the sheen or polish desired.

This Japanese gourd was treated with liquid floor wax to create a rich sheen.

Varnish

Many different types of wood finishes are available today, including varnish, shellac, and a wide range of plastic polymers such as varathane and polyurethane. They all come in a variety of finishes, from completely non-gloss matte to a very shiny surface. You may want to consider first staining the gourd with one of the many types of wood stains available in hobby and hardware stores. A combination stain sealer can also be used to completely seal the pores of the gourd shell before applying a varnish. All of the products that are formulated for use on wood, leather, or paper are suitable for use on the gourd surface.

Ornaments

Small gourds, including the mini-hardshell and all ornamental gourds, are frequently used for ornaments or holiday decorations. We are all familiar with the colorful fresh ornamental gourds in fall decorations, where they provide a bright color accent to any arrangement. Once dry, however, the skin is scraped away, and the gourd shell underneath those once-bright colors is a light tan. Use paints, stains or leather dyes to recreate those colors, or to make new decorations to fit with any holiday or home decor. Wax or varnish them to protect the surface, and you will enjoy them for years to come.

Many native people throughout North and South America would fasten small gourds together in a string which assumed magical qualities. These charm strings or spirit strings are still found in some villages or are part of rituals which evoke powerful natural forces. While the original charm strings were not ornamented, crafters today frequently embellish the small gourds with small designs, stencils, patterns or colors. The gourds can then be secured to a rope or ribbon chain, either by their stems, or by fasteners which are screwed or glued onto the gourd shell. Let your imagination be your guide to create a unique and lovely addition to your home or patio decor.

Small gourds make wonderful decorations for the holidays.

Make a charm string by decorating several small gourds and hanging in your patio. (Artist: Ardith Willner)

Cut an opening with a small saw.

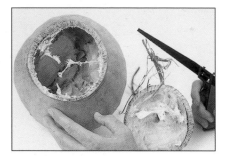

The interior is filled with dried pulp and seeds.

Scrape out with a spoon or scraping tool.

The photos above are by Jim Widess

Containers

Probably the most popular use of gourds through the millennia has been as a container. The various sizes and shapes of the gourds made them useful for myriad functions in the daily lives of people around the world.

While in other ages people used sharpened shells, stones or bone to cut the shell, many other choices are available today. While any saw will cut a gourd, a keyhole or similar narrow blade will allow you to cut along irregular shapes and lines. Saw blades that fit the Xacto hobby carving tools allow cuts along extremely delicate lines. First puncture the gourd shell using a kitchen knife or similar sharp blade. Then insert the saw blade and carefully saw along the line which has been marked with chalk or pencil. If you have trouble holding the gourd steady, try bracing it on a foam pad on your lap, or against the corner of a counter wall.

Once you have cut completely around the gourd, remove the top. Inside are many seeds and lots of dried pulp. If the gourd is still green, carefully scrape out some of the pulp and seeds, and fill the gourd with water. After a week or so, the remaining pulp will be mushy and easy to remove. Once the gourd is cleaned out, set it aside until the shell is entirely dry, a process that will probably take up to a month.

If you are working with a dry gourd, use hands and a scraping tool to clean out the interior completely. Poke around in the kitchen and garage to find tools that may be used for this job. (I find that a grapefruit spoon and a serrated-edge serving spoon are my favorite tools for this job. I usually use a metal kitchen scraping pad to scrape out the last bit of pulp from the shell, and to smooth the inside surface.) Rough sandpaper, particularly the type that is bonded to a sponge, is very useful for this task.

At this stage, you have many choices for further finishing your container.

For most purposes, the interior does not need to be finished further. If it is important to seal the surface, it can be coated with varnish, paint or melted wax. This will make the container relatively water-tight, so that it can be used to hold flowers or plants. (I would recommend that even though the interior surface has been sealed, a container such as a cut-off plastic drink bottle be used to hold water for flowers or a small plastic dish be used under potted plants.)

The exterior can be further embellished with paints, stains, dyes or varnish. Craft books mentioned in the bibliography will describe many other alternatives for you to consider. Gourd containers, plain and fancy, have been used throughout the world by almost every culture. As you make your own container, you are following a tradition that has been the part of daily life of people through the ages.

*A fancy gourd can hold an elegant
flower arrangement.
(Gourd grown by Jim Story;
arrangement by Davino Florists)*

*Use a plastic saucer
inside a gourd for a plant holder*

A SEASON *in the* GOURD GARDEN

Prepare soil with manure and fertilizers as soon as it can be worked.

DAY 1

Plant seeds when soil temperature reaches 70 degrees. (If you have a short growing season, start seeds indoors four weeks proior to date of last frost.)

3 WEEKS

Several leaves appear on plants, which seem to rest before vining.

12 WEEKS

Gourd vines cover fields with a sea of green leaves and sparkling white blossoms.

14 WEEKS

Mature green gourds nestle among the leaves. Do not pick them at this stage, however. They are still developing as long as the vines and leaves are green.

18 WEEKS

Allow the gourds to stay in the garden until the vines are dried and the stems are completely brown.

6 WEEKS

*Plants begin to vine.
If growing plants on a trellis, begin
to train at this time.*

8 WEEKS

*Male blossoms appear first
on the vines and female
blossoms appear soon after.
Hand pollinate plants to
insure a bountiful crop.*

10 WEEKS

*Baby gourds proliferate
in the garden.*

20 WEEKS

*You can leave the gourds in the
garden to dry, or bring them into
a well-venilated area.*

3-6 MONTHS
AFTER HARVESTING

*Allow to dry until the
gourd feels light and the seeds
rattle inside.*

*You can craft your gourds at any
time after they are completely dry.*

USDA Hardiness Zone Map

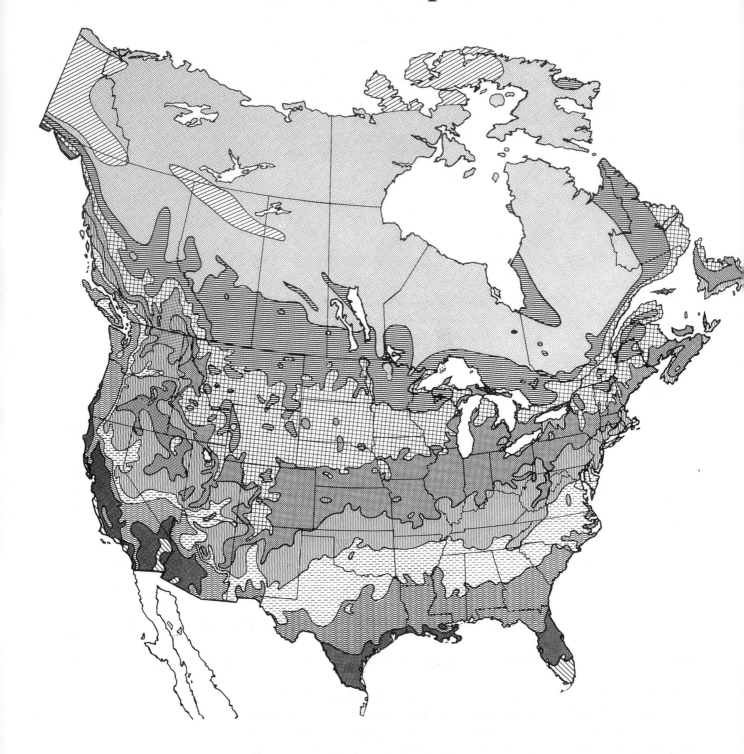

Zone Map Notes

Frequently in general purpose gardening books and articles that appear in the popular press, the author refers to gardening or plant "zones." The numbers range from 1 to 10 and refer to geographic areas that are based on data collected by the U.S. Department of Agriculture, specifically the average annual minimum temperature. Although these zone designations have become quite popular as a plant guide, they do not take into account other important data, such as the dates of first and last frost, average temperatures throughout the growing season, hours of sun, maximum temperatures, rainfall patterns, etc. For additional information on all of these factors, as well as weather data for specific regions, you are encouraged to contact the National Climatic Data Center, Asheville, NC 28801-5001 (phone: 704-271-4800. Request the bulletin that has Comparative Climatic Data for the United States through 1995.) Because their data is gathered from hundreds of weather stations located throughout the country, including Alaska, Hawaii and Puerto Rico, it reflects microclimates within broad geographic areas.

In addition to this source of information, many books which focus on gardening in regions (such as the *Sunset Western Garden Book*), provide their own zone designations which are completely independent of the USDA Hardiness Zones.

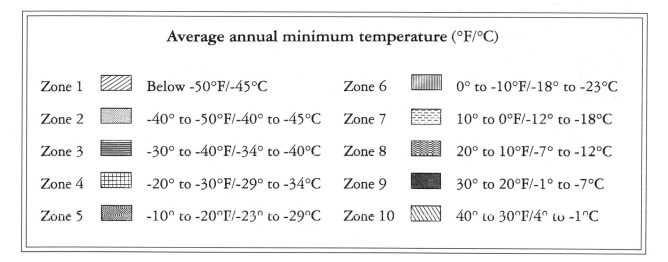

Average annual minimum temperature (°F/°C)

Zone 1	Below -50°F/-45°C	Zone 6	0° to -10°F/-18° to -23°C
Zone 2	-40° to -50°F/-40° to -45°C	Zone 7	10° to 0°F/-12° to -18°C
Zone 3	-30° to -40°F/-34° to -40°C	Zone 8	20° to 10°F/-7° to -12°C
Zone 4	-20° to -30°F/-29° to -34°C	Zone 9	30° to 20°F/-1° to -7°C
Zone 5	-10° to -20°F/-23° to -29°C	Zone 10	40° to 30°F/4° to -1°C

This map is based on data collected by the U.S. Department of Agriculture. However, it only reflects the annual minimum temperature, not temperature ranges at other times of the year. Other charts such as on page 28 – 29, and those found in the Farmers Almanac *or in* Sunset Western Garden Book *provide more specific information about growing seasons in selected areas in the U.S.*

The American Gourd Society

The purpose of the American Gourd Society is: To promote horticultural and ethnological research in gourds; to publish books, pamphlets bulletins; to encourage the use of gourds in decorative art; to hold exhibits.

The AGS has its roots in organizations that were first started in 1934 in California and in 1937 in Massachusetts. Currently, it is based in Mt. Gilead, Ohio, and boasts a growing membership of over 5,000, with representation in all states and many foreign countries.

While based in New England, AGS sponsored and published several scholarly monographs on gourd history and culture that continue to be a valuable source of information and education today. In addition, AGS has published several other booklets and bulletins that describe horticulture and simple craft projects. AGS also publishes the quarterly Gourd newsletter, which includes a list of growers who sell seeds and gourds of all types.

Several of the Gourd Society chapters sponsor annual gourd shows. The World's Largest Gourd Show, sponsored by the Ohio chapter, is held annually the first weekend in October in Mt. Gilead, Ohio. It attracts participants from around the country as well as international visitors. Many state chapters also sponsor annual gourd shows, featuring dried gourds and seeds, as well as many categories of crafts.

The states with active chapters of the American Gourd Society are North Carolina, Ohio, Virginia, Indiana, Kentucky, Texas, Florida, and California. Plans are underway for additional chapters in Arkansas, Connecticut, Illinois, Louisiana, Missouri and Pennsylvania.

Contact the American Gourd Society
for the name and address of your local chapter.

American Gourd Society
P.O. Box 274,
Mt. Gilead,
Ohio 43338-0274.
Phone: (419) 362-6446.

Other Members of the Cucurbitaceae Family

Buffalo Gourd, also known as Coyote Gourd (*Cucurbita foetidissima*)

This gourd was named because of the very strong smell that is released when even just the leaves are disturbed and it is part of the very large Cucurbita genus. It grows in dry, semi-arid areas in the southwestern United States and Mexico and was probably a native plant in this hemisphere. The plant is unlike the other gourds we have discussed, since it is a perennial. It has a very large root system, which has made it the object of intense research in recent years; scientists are exploring ways this plant can be grown in other desert areas where firewood is in short supply. The roots of this plant grow so rapidly and with such mass that within a very few years the plant can be harvested to provide a slow-burning, clean fire. The root also contains an abundance of starch, and the plant is being studied for commercial applications of this property.

Many other parts of the buffalo gourd have been useful to humans as well. Both the root and the fruit contains a saponin that produces suds; soap made from this source is used for laundry soap and shampoo.

The plant was also widely used in medicine by native people in this area. A poultice was made from the leaves and applied to open sores on humans and animals. The seeds of the buffalo gourd contain a large amount of oil and have a relatively high protein count, comparable to that of soybeans and peanuts. Historically the seeds were ground and eaten as a mush. Current research is investigating ways to extract the oils for both food and industrial uses. The gourd itself is small and round, approximately the size of a tennis ball. The shell is relatively thin, so although it will dry with a hard shell, it is not useful as a container.

Bitter Gourds, including Balsam Apple, Balsam Pear, Bitter Cucumber

These plants belong to the genus *Momordica*. The name refers to the odd shaped jagged edged seeds of the several members of the genus. The most familiar species is the balsam pear, which grows on a prolific vine that is frequently a nuisance weed which invades and overwhelms native shrubs and trees in the Southeastern United States. The leaves are small and deep lobed, hiding delicate soft yellow blossoms. The crescent shaped fruit grows to a length of 4-6 inches and turns a brilliant orange when it is mature, splitting open to expose several bright red seeds. The young fruit is used for food, particularly in Asia. Eaten fresh, it is often boiled or fried; frequently it is preserved in a pickling spice or brine and served either as a relish or as an ingredient in other dishes. The plant is more often used as a folk medicine, where the fruit is used to treat many ailments such as wounds, burns and itches; internally it is thought effective against rheumatism, coughs and menstrual cramps. The active ingredient which is contained in this species is similar to insulin, so any uses of the plant should be made with caution. The leaves are boiled to make a tea that is also reputed to have great healing powers.

The balsam apple is less common, but has also been used widely both for food and medical purposes, especially in eastern Asia.

Wax Gourd, also known as Chinese Preserving Melon, Tung Kwa, Mo Kwa, Cham Kwa, Mao Gwa, Chinese Fuzzy Gourd (*Benincasa hispida*)

This plant is native to Java, but has been widely cultivated in Japan and China. The name "fuzzy" refers to the unusual characteristic of fuzz that covers the vine and young plants. Grown primarily for food, several varieties are specifically picked when they are quite young, when they are served in a variety of preparations throughout Asia and India. The mature fruit can grow to be very large, up to 30 lbs., with a white waxy covering. It can last up to two years if stored in a cool, dark space. The flesh of the large wax gourd may be pickled or steamed, or the entire gourd can be stuffed with meats and other vegetables and steamed, much as we might prepare varieties of squash.

The wax gourd is considered an emblem of fertility, and frequently was presented to newly-married couples in India.

Snake Gourd

While there is a variety of Lagenaria gourd that is often called "snake gourd," the ones more often referred to by this name are natives of Asia, of the genus *Tricosanthes*. There are several species in this genus, but the one that is widely cultivated in China is *Trichosanthes cucumerina* (cucumber-like) which often attains lengths up to six feet. It is used as a food, most often being sliced, boiled or added to curries. The distinctive feature of this plant is the beautiful flowers; they are up to 2" across, with the petals surrounded by a lacelike fringe. They open late at night and emit a sweet but powerful fragrance, only to fade by morning.

Teasel Gourd (*Cucumis dipsaceus*)

This gourd is part of the same genus as most melons and the cucumber. It is native to Africa and Arabia, where it is frequently eaten, although it is rarely eaten in other areas. The distinctive feature of this plant is the dense covering of soft spines which protects the fruit, which is about the same size and shape as a kiwi.

Malabar Gourd, fig-leaf gourd, Siamese or Angora gourd. (*Cucurbita ficifolia*)

Although this plant is widely grown and eaten in Asian countries, it is native to Mexico, where it is also commonly raised for food. It is one of the earliest fruits known in the Western Hemisphere, where remains were identified in the Huaca Prieta site in Peru, dating from 5000 B.P. The fruit resembles a small watermelon, growing up to 14 inches long. When mature it is rich green with dotted white stripes running lengthwise. While it can be dried and the thin wooden shell used for crafts, the fruit is more frequently used for food. The flesh can be prepared in many ways as a vegetable, and the seeds are frequently roasted.

Chayote (*Sechium edule*)

This gourd originated in Mexico and Central America, where it was cultivated by the Aztecs. The fruit is pear shaped, either smooth or covered with soft spines, containing a single large flat seed. The fruit can be baked, boiled or stir-fried, as in Chinese dishes. The vine's large, fleshy, tuberous roots are also edible. The tender young shoots may be eaten, much as asparagus. This plant differs from other gourd plants we have examined, in that the entire fruit is planted, laid on its side, with the narrow end protruding. It grows very fast in warm, fertile soil, preferring tropical climates.

Turk's Turban

This lovely squash is a variety of the *Cucurbita maxima*, and therefore a close relative to the familiar pumpkin. It is frequently found in markets in the fall, and the unusual shape and brilliant orange colors make it a favorite in home decor. If kept in a cool place, the squash will last up to two years, although it will never dry with a hard shell. It can be eaten; as other winter squash it can be baked, boiled, stuffed or made into a soup or stew.

Suppliers of Seeds

American Gourd Society
Box 274
Mt. Gilead, Ohio 43338-0274

Write or call for complete list of variety of hardshells and ornamentals. Reliable and consistent germination.

Ms. Suzanne Ashworth
5007 Del Rio Rd.
Sacramento, CA 95822-2514
916-441-5036

A huge variety of hardshell gourd seed Excellent reliability.

Rocky Ford Gourds
PO. Box 222
Cygnet, Ohio 43413

Another superb source for gourd seeds of all varieties, reliable true-to-type.

The Gourd Garden and Curiosity Shop
4808 E. County Rd 30-A
Santa Rosa Beach, FL 32459
904-231-2007

A wide variety of hardshell and ornamental gourd seed, as well as dried and decorated gourds

Nichols Garden Center
1190 North Pacific Highway
Albany, OR 97321-4580
541-928-9280

Hardshell and ornamental seeds of many varieties; luffa (round and cylindrical)

Burpee
300 Park Ave
Warminster, PA 18991-0001
1-800-487-5530

Big Gourds Mix, large bottle, small fancy, crown of thorns, luffa

Lake Valley Seed
5717 Arapahoe
Boulder, CO 80303

Seeds for hardshells and ornamentals available through your local garden center.

Jung Seed and Nursery
335 S. High St.
Randolph, WI 53957-0001
1-800-692-5864

Two sizes of ornamental gourds
(mixed large and mixed small)
crown-of-thorn, bird house hardshells
turk's turban

Gurney's Seed and Nursery Co.
110 Capital St.
Yankton, SD 57079
605-665-9718

Large fruited mix, large and small
Decorative mix: Crown of Thorns, Spoon,
Birdhouse, Dipper, Balsam Pear,
Turk's Turban

Shepherd's Garden Seeds
30 Irene St.
Torrington, CA 06790
860-482-3638

Ornamental gourds; luffa
(This is a wonderful catalog to be
aware of. Hopefully they may carry
additional gourd seeds in the future.)

Pinetree Garden Seeds
Box 300
New Gloucester, ME 04260
207-926-3400

Italian cucuzzi, small spoon, small
warted, nest egg, crown of thorns,
birdhouse, long handled dipper,
corsican, luffa

DeGiorgi Seed Co.
6011 "N" St
Omaha, NE 68117-1634
1-800-858-2580

Birdhouse, bushel, penguin, bottle,
dipper, luffa, mixed ornamental,
Turk's Turban

Thompson & Morgan
Box 1308
Jackson, NJ 08527-0308
1-800-274-7333

Collins giant long gourd, Choose-
your-weapon mix, luffa, bottle, small
mixed, Turk's Turban

Seeds Blum
Idaho City Stage
Boise, Idaho 83706

No hardshell gourds, but a wonderful
assortment of gourd relatives: angled
luffa, melons (balsam apple, pear)
chayote, Chinese fuzzy melon (wax
gourds,) dinosaur gourd

Seeds of Change
Box 15700
Santa Fe, NM 87506-5700
1-888-762-7333

Birdnest, bottle, Hopi rattle, luffa

Suppliers of Dried Gourds

Many dried gourds are available at local farmers' markets or at small specialty farms throughout the country. For a more complete listing of gourd farms that ship gourds throughout the country, consult the *Gourd* newsletter of the American Gourd Society.

ALABAMA

Glenn Burkhalter
153 Wiljoy Rd.,
Lacey's Spring, AL 35754

Jim Cahela
257 Hipp Rd.
Blountsville, AL 35031

Palisades Gourd Farm
P.O. Box 591
Oneonta, AL 35121

Billy G. Levins
101 Gum Springs Cut-Off Rd.
Hartselle, AL 35640-9455

ARKANSAS

Dalton Farms
610 CR 336
Piggott, AR 72454

CALIFORNIA

Gourd Factory
Box 9
Linden, CA 95236

Gourd and Christmas Tree Farm
44103 Gadsden
Lancaster, CA 93534

Janice Monypeny
42226 Sultan Ave
Palm Desert, CA 92211

Pumpkin and Gourd Farm
101 Creston Rd
Paso Robles, CA 93446

Welburn Gourd Farm
40787 DeLuz Murrieta Rd.
Fallbrook, CA 92028

Zittel's Gourd Farm
6781 Oak Ave.
Folsom, CA 95630

FLORIDA

Gourd Garden and Curiosity Shop
4808 E. Country R 30-A
Santa Rosa Beach, FL 32459

GEORGIA

Lena Braswell Gourd Farm
Rte. 1, Box 73
Wrens, GA 30833

INDIANA

Sandlady's Gourd Farm
RR4, Box 86
Tangier, IN 47952

KENTUCKY

Gourd-Geous Creation of Kentucky
6635 Old Bloomfield Rd.
Bloomfield, KY 40008

MISSISSIPPI

Tom Keller
P.O. Box 1115
W. Point, MS 39773

MISSOURI

Ozark Country Creations
30226 Holly Rd.
Pierce City, MO 65723

NORTH CAROLINA

Fisher Farm
11901 Cooper Rd.
Nashville, NC 2

Terry Holdsclaw (Luffas only)
Box 85
Terrell, NC 28682

OHIO

Gourd Central
7264 SR 314
Mt. Gilead, OH 43338

PENNSYLVANIA

Betsy A. Rebuck
RD4, Box 218
Sanbury, PA 17801

Blessing Farms
475 Chapel Church Rd.
Red Lion, PA 7356

TENNESSEE

Glenda Wade
221 Frankie Lane
S. Fulton, TN 38257

TEXAS

West Mountain Gourd Farm
Rte. 1, Box 853
Gilmer, TX 75644

VIRGINIA

John Clark
3833 Bruce Rd.
Chesapeake, VA 23321

AUSTRALIA

John VanTol
Australian Gourdfather
Box 298
E. Maitland 2323, NSW Australia

Selected Bibliography

Ashworth, Suzanne. *Seed to Seed*. Seed Saver Publications, Decorah, Iowa 1991.

Bailey, L. H. *The Garden of Gourds*. Gourd Society of America, Inc. Mt. Gilead, Ohio. 1958.

Bird, Christopher O. *Modern Vegetable Gardening*. Lyons & Burford, Pub.
New York, New York, 10010. 1993.

Brenzel, Kathleen Norris, editor. *Sunset Western Garden Book*. Sunset Publishing Corp.
Menlo Park, CA. 1995.

Carr, Anna, Miranda Smith et. al. *Chemical-Free Yard and Garden*. Rodale Press,
Emmaus, PA. 1991.

Ellis, Barbara and Fern Bradley. *The Organic Gardener's Handbook of Natural Insect
and Disease Control*. Rodale Press, Emmaus, PA. 1996.

The Gourd. Quarterly Newsletter of the American Gourd Society. Mt. Gilead, Ohio. 1937-1997.

Gourd Society of America.*Gourds, Their Culture and Craft*. American Gourd Society,
Mt. Gilead, Ohio, 1966.

Heiser, Jr., Charles B. *The Gourd Book*. University of Oklahoma Press, Norman, Oklahoma. 1979.

Heiser, Jr., Charles B. *Of Plants and People*. University of Oklahoma Press,
Norman, Oklahoma. 1985.

Larkcom, Joy. *Oriental Vegetables: The Complete Guide for Garden and Kitchen*.
Kodansha International Ltd. Tokyo, Japan. 1991.

McClure, Susan and Sally Roth. *Successful Organic Gardening Companion Planting*.
Rodale Press, Emmaus, Penn. 1994.

Summit, Ginger and Jim Widess. *The Complete Book of Gourd Craft*, Lark Books,
Asheville, North Carolina. 1995

VanTol, John. *Gourd Grower's Handbook*. Australian Gourd Club,
Maitland, NSW, Australia. 1993.

Zitter, Thomas A., Donald L. Hopkins and Claude E. Thomas, editors.*Compendium of Cucurbit
Diseases*. American Phytopathological Society, St. Paul, Minn. 1996.

Index

Your Garden Notes

Hillway Press
ORDER FORM

FAX ORDERS: 415-941-1613

MAIL ORDERS: HILLWAY PRESS
                P.O. BOX 592
                LOS ALTOS, CA 94022

Please send me _____ copies of **GOURDS IN YOUR GARDEN** @ $19.95 ea.   _____

Also by the same author: *The Complete Book of Gourd Craft*
                          by GInger Summit and Jim Widess

Please send me _____ copies of

THE COMPLETE BOOK OF GOURD CRAFT @ $26.95 ea.   _____

Sales tax: add 7.75% for books shipped to a California address   _____

Shipping and handling $4.00

TOTAL   _____

METHOD OF PAYMENT

☐ Check   ☐ Visa   ☐ Mastercard

Name _____

Address _____

City _____ State _____ Zip _____

Telephone _____